A Concise Manual of

Engineering Thermodynamics

A Concise Manual of

Engineering Thermodynamics

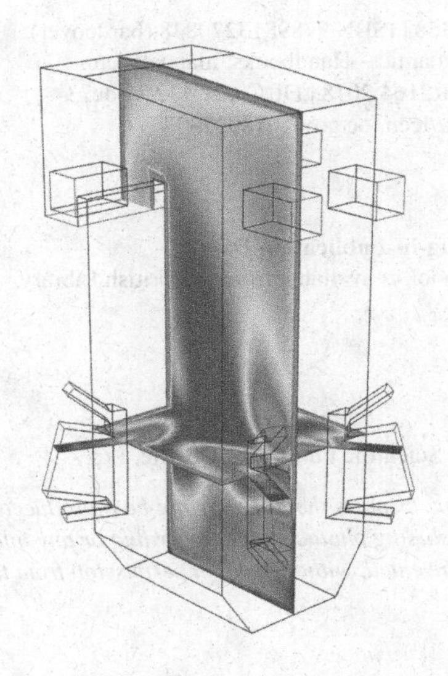

Liviu F Radulescu
Conestoga College, Canada

World Scientific

NEW JERSEY • LONDON • SINGAPORE • BEIJING • SHANGHAI • HONG KONG • TAIPEI • CHENNAI • TOKYO

Published by

World Scientific Publishing Co. Pte. Ltd.
5 Toh Tuck Link, Singapore 596224
USA office: 27 Warren Street, Suite 401-402, Hackensack, NJ 07601
UK office: 57 Shelton Street, Covent Garden, London WC2H 9HE

Library of Congress Cataloging-in-Publication Data
Names: Radulescu, Liviu F., author.
Title: A concise manual of engineering thermodynamics / Liviu F. Radulescu, Conestoga College, Canada.
Description: New Jersey : World Scientific, [2018] | Includes bibliographical references and index.
Identifiers: LCCN 2018027358 | ISBN 9789813270848 (hardcover)
Subjects: LCSH: Thermodynamics--Handbooks, manuals, etc.
Classification: LCC TJ265 .R2164 2018 | DDC 621.402/1--dc23
LC record available at https://lccn.loc.gov/2018027358

British Library Cataloguing-in-Publication Data
A catalogue record for this book is available from the British Library.

For any available supplementary material, please visit
https://www.worldscientific.com/worldscibooks/10.1142/11002#t=suppl

Desk Editor: Tay Yu Shan

Typeset by Stallion Press
Email: enquiries@stallionpress.com

Printed in Singapore

Were I not a beacon but a candle, that would be enough. Were I not a candle either, that would still suffice as I have striven to bring light into darkness.

Nicolae Titulescu — Romanian diplomat (1882–1941)

Preface

The specific requirements of Mechanical Systems Engineering (MSE) program at Conestoga College in Kitchener (Ontario) revealed the need for a customized textbook, to cover in enough detail the unit outcomes as described in the outline for the *Thermodynamics* course. However, the writing of a textbook is a very demanding and time-consuming enterprise. The book you are holding in your hands — covering the fundamentals of applied thermodynamics including the heat transfer — is intended for the students working toward earning their Bachelor of Engineering degree.

The engineering field, especially in its mechanical branch, requires a good understanding of fundamental concepts including thermodynamics. Mechanical engineers use these concepts to design and analyze a variety of industrial equipment and machinery, many of them making use of heat as a primary source of energy, as in the case of power plants, heating and cooling systems, motorized vehicles, aircraft, ships and many more. Engineers from other branches — civil or electrical — need to have knowledge of the basic thermal phenomena, as they need working knowledge of physics, mathematics and chemistry. This textbook aims at offering engineering students not only a strong theoretical foundation, but also knowledge of applications centered on energy conversion, heat transfer and environmental control. This is the reason for the title "Engineering Thermodynamics".

For measuring units, priority has been given to SI, due to its near-worldwide adoption.

This edition includes a collection of some 50 carefully tailored homework problems together with solutions covering all 13 chapters, a number of questions with answers to promote a greater understanding of the subject, as well as relevant property tables and diagrams. This is important because what I recommend to all my students is to solve as many problems as possible. There is no better way of understanding engineering thermodynamics than by doing exercises. I am confident that the interested student will learn quickly and easily from these problems.

I would like to express my gratitude to Dr. Paul-Dan Oprişa-Stănescu of the Politehnica University of Timişoara, Romania, for his valuable contributions, his critical review of the original manuscript, and for his special attention to accuracy.

I would like to thank the team of editors from World Scientific, especially Ms. Yu Shan Tay and Mr. Taisuke Soda, for their guidance and help in the preparation of the manuscript for publication.

It is my hope that this entire book will serve as an introduction for a better understanding and appreciation of industrial thermal equipment as well as for more advanced study.

Liviu F. Radulescu
Kitchener, 2018

Problem Solving Guide

Thermodynamics problems are relatively easy to solve when a methodical procedure is applied. Practically any thermodynamics problem can be approached following the steps described below:

1. Sketch the thermodynamic system and show the energy interactions across the boundaries.
2. State necessary assumptions (e.g., the working substance is an ideal gas, the process is reversible, etc.)
3. Determine the property relation (e.g., the ideal gas equation of state). Establish what properties are known.
4. Determine the process and sketch the process diagram. Determine what other properties can be calculated.
5. Take into account physical constraints imposed by the problem statement, such as the volume is halved, or the pressure is doubled during the process.
6. Apply conservation of mass and conservation of energy principles.
7. Develop enough equations for the unknowns and solve for desired quantity, first algebraically then substituting numbers into equation.
8. Do a sanity check for the magnitude of the answer to see if the solution "makes sense".

Note

Problems are found at the end of each chapter. Answers have been obtained using the tables of properties and diagrams found in Appendix E. Constant specific heats have been used in the solutions. Calculations have been performed using Microsoft Excel and the final results have been rounded to a reasonable level of accuracy. In the conversion of temperatures between degrees Celsius and kelvin, all calculations use the rounded value of 273 (instead of the exact 273.15): T (K) = t (°C) + 273.

Contents

Chapter 1

Basic Concepts of Thermodynamics

1.1. Introduction

Thermodynamics is a part of physics that studies the relationship between the properties of substances and the quantities we refer to as "work" and "heat" when the substances in question undergo a change of state [1, p. 3]. The term "thermodynamics" (initially spelled "thermo-dynamics" comes from two old Greek (Gr.) words *θέρμός* [*thermos*], meaning "hot" and *δυναμικός* [*dunamikos*], meaning "powerful, strong" [2] which suggests the fact that the object of study is the thermal energy, which can be transferred from one substance to another and converted to other forms, hence its "power" (see also [3]). Considering the corpuscular structure of matter, the **thermal energy** can be defined as the part of the overall energy[a] of a system that comes from the movement of atoms and molecules in matter. It is a form of kinetic energy associated to the random translational movement of those particles. The thermal energy of a system can be increased or decreased. If one adds to this thermal energy (kinetic energy) the potential energy associated with the binding forces that hold the particles together, one obtains the **internal energy**. However, in order to study thermal phenomena, it is not necessary to make reference to the corpuscular

---

[a]Energy is a description of a quality or a condition, representing the capacity to produce a change.

structure of matter. This is what the so-called **phenomenological thermodynamics** (also named classical thermodynamics) does. It emerged as a science in the seventeenth century, and most of its fundamental laws were formulated before 1870, way before the modern atomic theory. The classical thermodynamics represents a generalization of extensive empirical evidence, and its conclusions are supported by the **statistical thermodynamics** (or statistical mechanics), emerged in the late nineteenth century, which explains thermal macroscopic phenomena through statistics and mechanics (classical and quantum) at the microscopic level. **Engineering thermodynamics**, studying the conversion, transfer and use of thermal energy in engineering applications, is essentially phenomenological thermodynamics.

1.2. Quantities measuring units and dimensions

Because engineering thermodynamics is mainly an experimental science, measurements play an important role. The process of measuring involves a comparison of a certain quantity to a measuring unit of the same type. The result of a measurement is a real number (the ratio of the magnitude of the quantity to the magnitude of the unit) times the measuring unit. Quantitative data in the real world being analog (continuous), not digital (discrete), the result of any measurement is not an exact value, only an approximation. Consequently, the result of any measurement will be affected by errors. Two important terms are defined in connection to the errors of measurement: precision and accuracy.

Precision refers to the smallest portion of a unit to which a measurement is taken. For example, 12.3 m is precise to a 1/10 of a meter. The number of **significant digits** here is 3 (1, 2 and 3). The digits (or figures) in the result of a measurement are called "significant" because they have a certain significance with respect to precision and accuracy [4, p. 30]. Nonzero digits are always significant. The case of zeros is more complex. For instance, zeros in front of other digits are not significant (0.045 has two significant digits), but trailing zeros to the right of a decimal are significant

(6.70 has three significant digits). Trailing zeros in integer numbers are ambiguous when assessing the level of significance; therefore, to avoid uncertainty, use the scientific notation to place significant zeros behind a decimal point. The number of decimal digits shows implicitly the precision of measurement.

Accuracy refers to the number of digits in a measurement that are known to be accurate, or true, or significant. For example, 0.0123 m is accurate to three figures. The number of significant digits here is 3, which reflects also the accuracy of the measurement.

When performing calculations using measured values, keep in mind that no result of a calculation can be more accurate than the least accurate of the quantities entered. As a rule of thumb, perform calculations with a precision of at least $(n + 1)$ significant digits and round only the final result to the required number of significant digits, n.

There are several systems of weights and measure currently in use, but the system that is nearly globally adopted is the **International System of Units** (abbreviated **SI**, from the French "le système international d'unités"), developed in 1960 from the meter–kilogram–second (mks) system [5].

SI (often referred to as "metric"), largely used in physics and engineering, is centered on seven **base units**, defined in an absolute way, corresponding to seven base quantities: meter [m] for length; kilogram [kg] for mass; second [s] for time; ampere [A] for electric current; kelvin [K] for thermodynamic temperature; mole [mol] for amount of substance; candela [cd] for luminous intensity.

From these seven base units several other units are derived. Each **derived unit** is defined purely in terms of a relationship with other units; for example, the unit of velocity is 1 m/s (i.e., the measuring unit for length over the measuring unit for time).

Units accepted for use with the International System are: minute [min], hour [h], day [d], degree of arc [°], minute of arc [′], second of arc [″], liter [l or L], tonne [t].

A prefix may be added to units to produce a multiple of the original unit. All multiples are integer powers of ten. The greatest advantage of the SI consists of its coherence, in the sense that when only base and derived units are used, conversion factors between units are never required.

Still in use in North America, especially in the United States, is the **United States Customary System** (also called American system or, sometimes, "English units") developed from the British imperial system. Since in the US Customary System the multiples of original units are not powers of ten, most engineering calculations require the use of **conversion factors**, even within the system.

However, a physical law must be independent of the units used to measure the physical variables. Based on this postulate, the **dimensional analysis** has been developed. It is a tool to check relations between physical quantities by using the concept of **dimension** [6, pp. 17–24]. By definition, each type of base quantity has its own dimension. For instance length L, mass M, time τ, and temperature T are base dimensions. The dimension of a derived quantity can be described using a combination of basic physical dimensions according to a physical equation; for example, mass flow rate has the dimension "M/τ", and may be measured in kilograms per second, pounds per minute, or other units. An immediate consequence is that any meaningful physical equation must have the same dimensions on both sides of the equals sign. This property is called **dimensional homogeneity** of a physical equation and checking it is the basic way of performing dimensional analysis. Another important use of the dimensional analysis is to determine the dimension of a variable in an equation. The dimension for the resultant quantity can be derived by simply substituting each quantity in the equation with its base dimension, omitting any numerical coefficients, and doing the algebra. For example, from the physical equation for kinetic energy $mV^2/2$, its derived dimension is

$$M(L/\tau)^2 = ML^2\tau^{-2}$$

or, using SI dimensional symbols (measuring units),

$$\mathrm{kg\ m^2\ s^{-2} = (kg\ m\ s^{-2})\ m = N\ m = J.}$$

Thus the principles of dimensional analysis led to the conclusion that meaningful physical laws must be expressed by homogeneous equations in their various units of measurement. This was ultimately formalized in the **Buckingham Pi theorem** [7, p. 297], providing a method for computing dimensionless parameters, like the **similitude criteria** used in the study of convective heat transfer.

1.3. Thermodynamic systems

For a correct analysis of thermodynamic processes, some basic concepts need to be defined first.

The term "system" was used already in Sec. 1.1 when talking about the overall energy. A **thermodynamic system** can be seen as the portion of the universe, with a finite volume, that makes the object of some investigation (see Fig. 1.1). The rest of the universe outside the system is known as the **surroundings** (or the environment). The system and the surroundings are separated by an imaginary layer called **boundary**.[b] The boundary enveloping the system, like a surface in geometry, has no thickness and does not belong to the system nor to the surroundings; depending on the

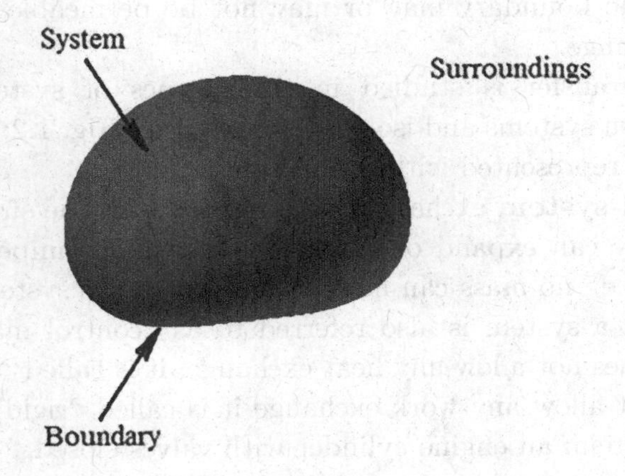

Fig. 1.1

[b]Sometimes the system is limited by a physical boundary, like the walls of a vessel; if the system is the content of the vessel, the boundary is the interface between the walls and the substance inside the vessel, while the walls themselves are considered part of the surroundings.

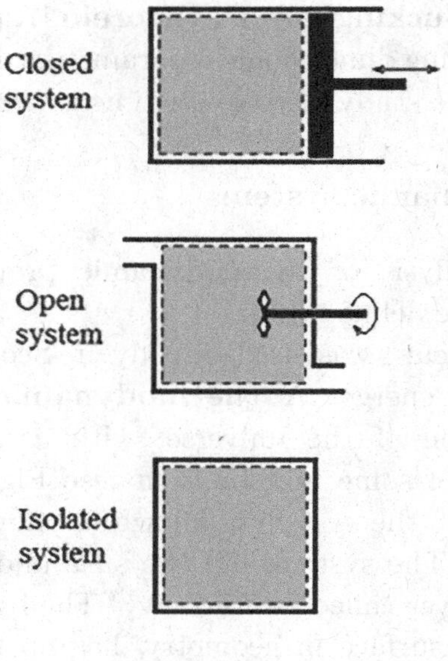

Fig. 1.2

situation, the boundary may or may not be permeable to mass or energy exchange.

Energy transfer is studied in three types of systems: closed systems, open systems and isolated systems (see Fig. 1.2; the system boundary is represented with dashed line).

A **closed system** exchanges only energy with the surroundings. Its boundary can expand or contract but remains impermeable to mass flux,[c] i.e., no mass can flow into or out of the system. For this reason, such a system is also referred to as "control mass". If the boundary does not allow any heat exchange, it is called "adiabatic"; if it does not allow any work exchange it is called "rigid". Example of closed system: an engine cylinder with valves closed.

An **open system** exchanges both mass and energy with the surroundings. Its boundary is permeable to mass and energy flux and has a fixed position, enclosing a constant volume. For this reason, such a system is also termed "control volume". When the

---

[c]**Flux** is defined as the amount that flows through a specified area per unit time [66].

flux of mass and energy transfer is constant with time, the conditions are described as "steady" (e.g., steady flow). Certainly, if an open system has its boundary impermeable to heat exchange, it is called "adiabatic". Example of open system: a steam or gas turbine.

An **isolated system** does not exchange mass or energy with the surroundings. A closed thermos bottle is essentially an example of isolated system (assuming that its insulation is perfect).

Since engineering thermodynamics studies only the macroscopic aspect of phenomena, any system is treated as a **continuum**, neglecting the fact that real substances are composed of discrete molecules with intermolecular spacing amongst them.

1.4. Properties of a system

In thermodynamics, the condition of a system is referred to as the (thermodynamic) **state** of the system which is defined by specifying values of a set of characteristics called thermodynamic **properties** (e.g., pressure p, volume V, density ρ, temperature T, mass m, internal energy U). Because they define the state of a system, thermodynamic properties are also referred to as **state functions, functions of state**, **state quantities**, or **state variables**. Some of these properties depend on the mass of the system (like V, m, U), and are called **extensive**; some others do not, and are called **intensive** (like p, ρ, and T). To quickly determine whether a property of a homogeneous system is intensive or extensive, think what happens if the system is divided in half: if the property in question remains unchanged, that property is intensive; otherwise it is extensive. If the extensive properties are referred to the unit mass, they become "specific quantities" and as such turn into intensive properties. For example, dividing the volume V of a system by its mass m one obtains the specific volume v; in a similar way, the specific internal energy $u = U/m$. Specific values are always symbolized by lower-case characters. Some of the properties are directly measurable, like pressure, volume and temperature. Other properties, like density, specific volume and others — as we will see in the next sections — can only be calculated. The main intensive properties of substances are presented in tabular form or, more intuitively, as **property diagrams** on thermodynamic coordinates.

The value of a property, in order to be representative for the state of the system, must be independent of the process through which the system has passed to reach that state. Also the intensive properties must have unique numerical values throughout the system [1, p. 7].

The state of a system in which properties have distinct, stable values throughout the system as long as external conditions are unchanged is called an **equilibrium state**. An isolated system reaches spontaneously equilibrium after a period of time. The process by which the system reaches equilibrium is called **relaxation** [8], and the time required for a system to reach equilibrium is called the **relaxation time**. Relaxation times can vary dramatically, from nanoseconds [9] — in the case of the pressure in gases — up to years — in the case of the concentration of solid solutions.

Since thermodynamic computations are performed using state functions and these functions are relevant for the state of a system only if the system is in perfect equilibrium or very close to it (quasi[d]-equilibrium), one can conclude that classical thermodynamics only describes the properties of various macroscopic systems at, or near, equilibrium [10].

In order to completely describe the state of a simple[e] system, we do not need to know the values of all its properties. It has been determined experimentally that a simple (closed) compressible system is completely defined by only two independent intensive properties [11]. This statement is known as the **state postulate**. As a direct consequence, any property can be expressed as a function of the two independent properties. Also, any equilibrium state can be represented as a two-coordinate point on a thermodynamic diagram ($p-v$, $p-T$, etc.)

[d] In Latin, *quasi* means "as if" or "almost".

[e] A system is considered to be a simple one in the absence of electromagnetic fields, surface tension, and macroscopic motion. This is a common situation to many engineering applications.

If x, y are two independent thermodynamic properties and z is another (dependent) thermodynamic property, then

$$z = f(x, y) \tag{1.1}$$

Equation (1.1) is known as an **equation of state**.

A **process** is the change a system undergoes from one equilibrium state to another. The series of states through which the system passes during a process is called **path**. A closed system is said to undergo a cyclic process or **cycle** when it passes through a series of states in such a way that (only) its final state is equal in all respects to its initial state [1]. Thermodynamics is about what processes are possible and what energy exchanges occur between a system and its environment during those processes. Also, we are often interested in graphically representing thermodynamic processes on thermodynamic coordinates. This is possible only if the path consists of equilibrium states, since properties define a state only when a system is in equilibrium. But real processes involve finite changes, produced by unbalanced forces; therefore the system will pass through nonequilibrium states, which cannot be dealt with in classical thermodynamics. Although it looks like we reached an impasse, a solution to this problem is possible, however, by considering that only infinitesimal unbalances exist at any given time, so that the process can be viewed as consisting of a series of quasi-equilibrium states. For this assumption to hold, the process must be slow, but not infinitely slow; just slow when compared to the relaxation time. For a gas (a phase dealt with most frequently in engineering thermodynamics), for example, this is almost always true, so it is a very good approximation to consider the thermodynamic processes as consisting of a succession of equilibrium states, which can be represented by full lines on charts [9].

If the thermodynamic state of a system changes from 1 to 2, then the change in the value of a state function z is independent of the path taken by the system when going from the initial state to the final state. The infinitesimal change of the state function is the differential of that function, dz, (also called *total* or *exact differential*) and the total change can be calculated by integration:

$$\int_1^2 dz = z_2 - z_1 = \Delta z \tag{1.2}$$

The reciprocal of this statement is also valid, that is if a quantity z is independent of path — i.e., it satisfies Eq. (1.2) — that quantity is a state function.

1.5. Zeroth law of thermodynamics

It has been determined experimentally that if two bodies are in thermal equilibrium with a third body, they are in thermal equilibrium with each other. This statement is called **the principle of transitivity of thermal equilibrium**, also known as **the zeroth law of thermodynamics**.[f] According to this law, the thermal equilibrium is a Euclidean relation. It is furthermore a relation of equivalence between an indefinite number of systems. Also, the zeroth law is the basis of thermometry. Thus, the law establishes an empirical parameter, called **temperature**, as a property of a system such that systems in equilibrium with each other have the same temperature. If we want to know if two bodies, A and B (see Fig. 1.3), are at the same temperature, it is not necessary to bring them into contact and watch whether their characteristics change with time. It suffices if a third body C, called **thermometer**, is brought into thermal equilibrium first with body A and then with body B by placing it in contact with A and B, respectively. The thermometer must possess an easily observable characteristic termed a thermometric property, e.g., the length of a column of mercury in a capillary tube. A scale of temperature must be predefined by assigning numbers to equally spaced divisions on the capillary tube. If the temperature readings on thermometer will be identical, the two bodies A and B are in thermal equilibrium. The zeroth law

[f]The principle was first formulated by the physicist Ralph Howard Fowler in 1931 [64, p. 40]. The term "zeroth law", appeared for the first time in 1939 in a book by Fowler and Guggenheim [65, p. 56], to show that, although formulated long after the other three laws of thermodynamics, its content is fundamental.

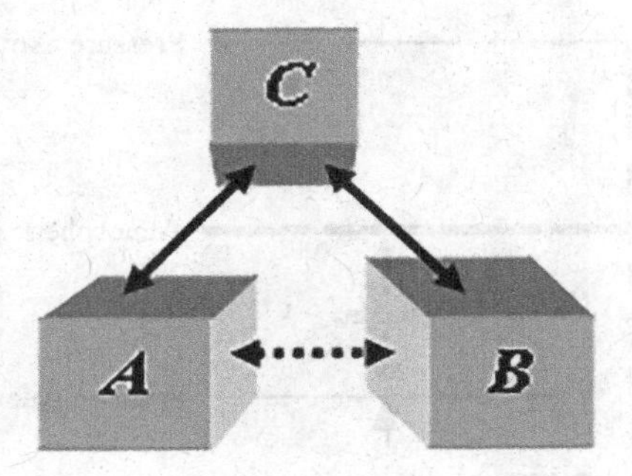

Fig. 1.3

can be restated as follows: "Two bodies are in thermal equilibrium if both have the same temperature reading even if they are not in contact" [11]. In practice, the thermometer should have a small body compared to A and B, otherwise the states of the two bodies will be altered during temperature measurement. The temperature determined using a thermometric property is also called **empirical temperature**, since the temperature scale used is arbitrarily defined. The only necessary condition for any thermometric property is reproducibility. The most common empirical scales of temperature are the **Celsius scale** (with °C as a unit) and the **Fahrenheit scale** (with °F as a unit). It will be shown later on that a temperature scale can be defined based on the second law of thermodynamics, without reference to any thermometric property. Such a scale, termed **thermodynamic scale**, is used to measure the **thermodynamic temperature** (with K [kelvin] as a unit in SI and °R [degree Rankine] in US Customary System). Thermodynamic temperature can be either zero (absolute zero) or positive. In this book we will use the symbol T for thermodynamic temperature and t for the empirical temperature. All thermodynamic equations involving temperature require the use of the thermodynamic scale.

Another principal parameter in thermodynamics is pressure, p. It too can only be zero or greater than zero, reason for which it is also called **absolute pressure** (see Fig. 1.4). In practice, most measuring devices used to measure pressures above the atmospheric level, named pressure gauges, indicate a value relative to the ambient air

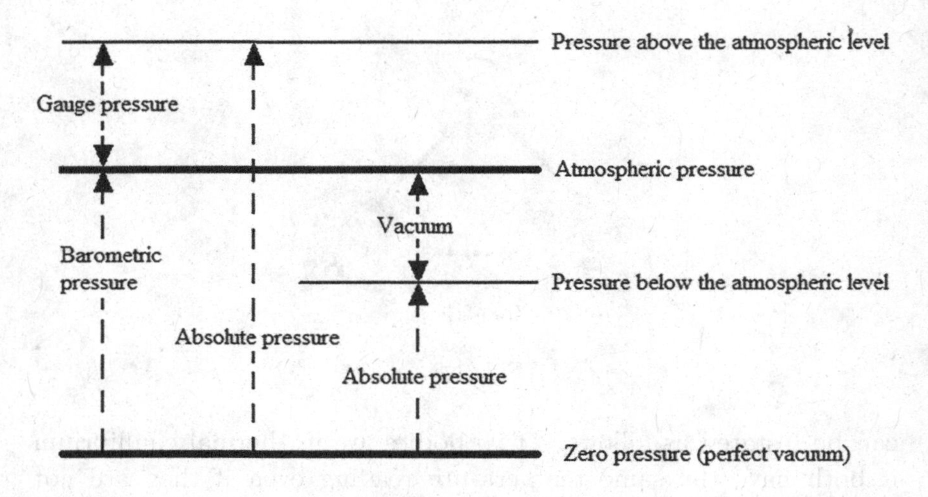

Fig. 1.4

pressure; this is called **gauge pressure**, p_g, and is equal to absolute pressure minus atmospheric pressure. The difference between the atmospheric pressure and an absolute pressure below the atmospheric level is termed **vacuum** and it is measured using vacuum gauges. All thermodynamic equations involving pressure require the use of the absolute pressure.

Basic concepts of thermodynamics — problems

1.1. A manometer connected to a pressure vessel indicates 2.8 bar. Knowing that the ambient pressure is 10^5 Pa, determine the gauge pressure p_g, and the absolute pressure p inside the vessel, in Pa.

1.2. What will be the indication of a Bourdon-tube gauge if the absolute pressure inside a container is 16 bar while the atmospheric pressure during measurement is 750 torr?

1.3. The temperature expressed in degrees Fahrenheit is an example of: (a) thermodynamic temperature; (b) relative temperature; (c) empirical temperature.

1.4. The turbocharger of an internal combustion engine can be considered as an example of (a) closed system; (b) open system; (c) isolated system.

1.5. Pressure, volume and temperature are: (a) state variables; (b) functions of state that cannot be directly measured; (c) state functions that can be measured directly.

1.6. The measurement of a thermodynamic property known as temperature is based on: (a) the zeroth law of thermodynamics; (b) the first postulate of thermodynamics; (c) the first law of thermodynamics.

Chapter 2

Properties of Pure Substances

In Secs. 1.3 and 1.4 we have been introduced to some assumptions used in classical thermodynamics: continuum, homogeneity, equilibrium. Another assumption is that we deal with **pure substances**; these substances have a fixed chemical composition throughout. This is very important when analyzing thermodynamic processes involving phase changes. A pure substance does not have to be a single chemical element; therefore water, oxygen, nitrogen, carbon dioxide are valid examples. Homogeneous mixtures of gases, such as air, can also be considered pure substances as long as there is no change of phase.[a]

2.1. Phases of a pure substance

A **phase** is a distinct arrangement of the molecules of a substance. Each phase presents an identifiable boundary surface. Pure substances can exist in three principal phases: solid, liquid and gas. (Plasma, a fourth phase, is not used in common engineering applications.)

Solids are characterized by the low energy levels and strong molecular bonds which limit the liberty of motion of atoms or

[a]Dry gaseous air contains approximately 78 % nitrogen, 21 % oxygen and 1 % other gases. When liquefying air, its composition will change during the process since nitrogen condenses at a cooler temperature than the oxygen and other constituents.

molecules to mere oscillations about a position of equilibrium. Consequently, solids have definite shape and volume.

Liquids are characterized by about the same molecular distances as the solids, intermediate levels of intermolecular bonds and energy. Consequently, liquids have definite volume but take the shape of a container.

Gases are characterized by large intermolecular distances, weak bonds and highest energy level. Consequently, gases have no definite shape and volume, tending to fill the entire volume of a container.

At various pressure–temperature combinations, any substance can be found in a gaseous, liquid or solid phase. For instance, at room temperature and pressure, copper is solid and water is liquid; at 1100 °C, copper is liquid and water is gas. As a pure substance undergoes a **phase transition** from solid to liquid and then to gas, the strength of its molecular bonds decreases and the internal energy level increases. Therefore the process of vaporization is associated with absorption of energy, while condensation is associated with release of energy.

In this course we will not consider the solid phase nor phase transitions involving solids, because engineering thermodynamics deals only with liquid-to-gas transition (or vice versa) to generate power.

2.2. Diagrams for phase-change processes

A graphical representation of what phases of a pure substance are present at any given temperature and pressure is called a **phase diagram**. Any substance presents a phase diagram (created experimentally). A typical example (not to scale) is shown in Fig. 2.1 (adapted from [12]).

In this p–T diagram, the three phases — solid, liquid and gas (vapor[b]) — are separated by three lines (see Fig. 2.1) and two special points marked C and T.

---

[b]There are subtle differences between the terms *vapor* and *gas*. If a substance is commonly a liquid at or around room temperature, its gaseous phase is usually called *vapor*. If a substance is found at room temperature above its critical point (like oxygen or hydrogen, for example), that substance will be described as a *gas*.

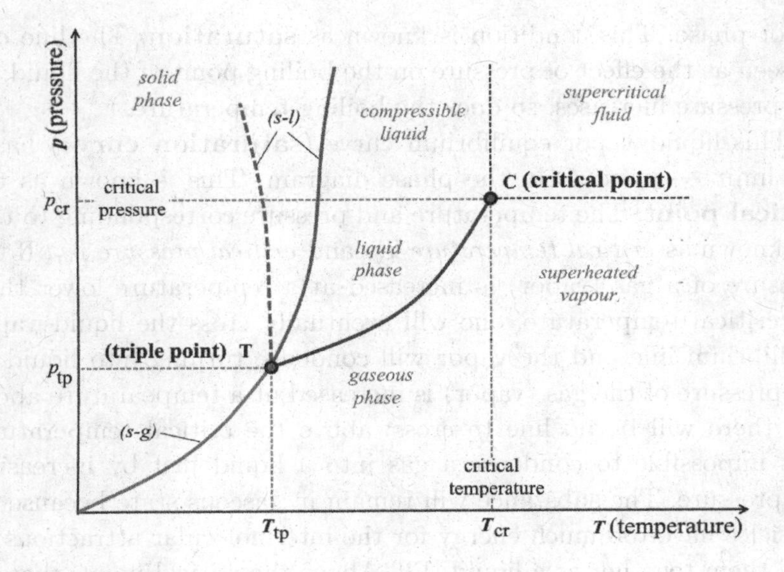

Fig. 2.1

The $(s-g)$ line separates the solid and gas (vapor) areas of the diagram. The phase transition from solid to vapor is known as *sublimation*; the reversed process is called *deposition*.

The $(s-l)$ line separates the solid and liquid areas of the diagram; it simply shows the effect of pressure on the melting point. For most pure substances this line has a positive slope (the continuous line); this shows that raising the pressure raises the melting point of most solids. For water, the solid–liquid equilibrium line (the dashed melting point line) has a negative slope; that is higher water pressures mean lower melting (freezing) points.

The $T-C$ line separates the liquid and gas (vapor) areas of the diagram; one can move from liquid to vapor easily either by changing the temperature (at constant pressure p = cst) or the pressure (at constant temperature, T = cst). The former is more common since in most engineering applications involving phase transitions processes occur at constant pressure (if we neglect pressure losses due to friction). The liquid turns to vapor — it boils — when one crosses the boundary line between the two areas. Anywhere along this line, there will be an equilibrium between the liquid and the vapor phase. This condition is known as **saturation**. The line can be seen as the effect of pressure on the boiling point of the liquid: as the pressure increases, so does the boiling temperature.

This liquid–vapor equilibrium curve (**saturation curve**) has a top limit — point C in the phase diagram. This is known as the **critical point**. The temperature and pressure corresponding to this are known as *critical temperature* T_{cr} and *critical pressure* p_{cr}. If the pressure of a gas (vapor) is increased at a temperature lower than the critical temperature, one will eventually cross the liquid–vapor equilibrium line and the vapor will condense turning into liquid. If the pressure of the gas (vapor) is increased at a tempearature above T_{cr}, there will be no line to cross: above the critical temperature, it is impossible to condense a gas into a liquid just by increasing the pressure. The substance will remain in gaseous state because its particles have too much energy for the intermolecular attractions to hold them together as a liquid [13]. Above the critical point, there is no clear distinction between the liquid and vapor phases, therefore no identifiable boundary can be noticed between compressible liquid and supercritical fluid.

At the opposite end of the liquid–vapor equilibrium curve there is point T, called the **triple point**, where all three phases (solid, liquid and gas) are in equilibrium together. That is why it is called a triple point.

Any (pure) substance has a triple point and a critical point, as one can see from the example in Fig. 2.2 [14]. Phase diagram (a) is for water and (b) for dry ice (CO_2). (Triple points are: A for water and X for CO_2. Critical points are: D for water and Z for CO_2.)

It is important to keep in mind that for any pure substance undergoing a phase change the pressure and temperature are not independent variables. There is a biunivocal relationship between pressure and temperature at saturation: to each saturation temperature corresponds a unique saturation pressure and vice versa, such that the higher the pressure, the higher the saturation temperature. This can be seen in Fig. 2.3 (the vapor pressure curve).

This relationship (that can be obtained experimentally) is very important for applications involving working fluids like water and

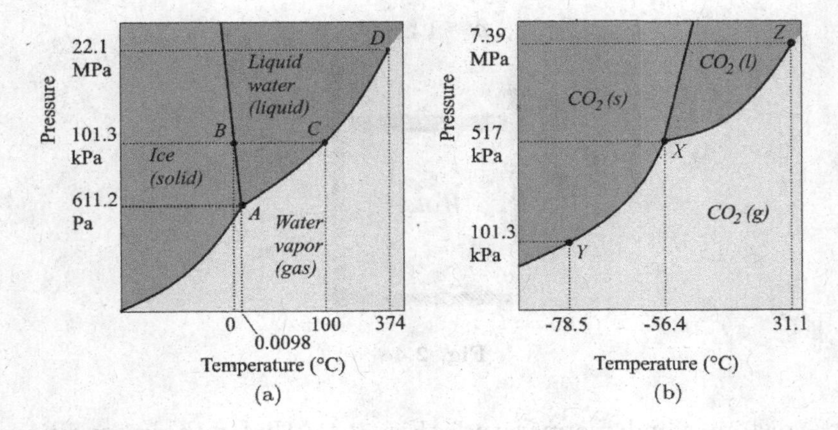

Fig. 2.2

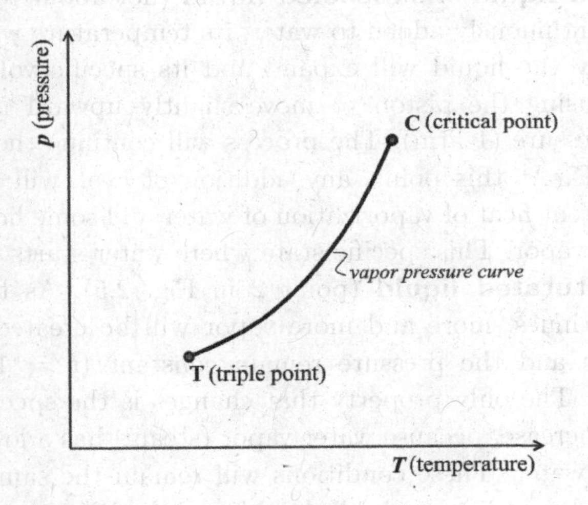

Fig. 2.3

refrigerants used in steam power plants and refrigeration units, respectively.

The liquid-to-vapor phase transition can be studied experimentally and the results of measurements can be recorded in a $T-v$ diagram. Let us assume that we have a quantity of water at room temperature (about 20 °C) and normal atmospheric pressure (1 atm)

Fig. 2.4

in a piston–cylinder apparatus[c] (Fig. 2.4). Under these conditions (point 1 in Fig. 2.5) water is in liquid phase and it is called **compressed liquid** or **subcooled liquid** (not about to vaporize). If heat is continuously added to water, its temperature will increase; consequently the liquid will expand and its specific volume v will increase causing the piston to move slightly upward maintaining constant pressure (1 atm). The process will continue the same way until 100 °C. At this point, any addition of heat will be used to cover the latent heat of vaporization of water and some liquid will be turned into vapor. This specific state where water starts to vaporize is called **saturated liquid** (point 2 in Fig. 2.5). As the heating process continues, more and more vapor will be created while the temperature and the pressure remain constant (t = 100 °C and p = 1 atm). The only property that changes is the specific volume which will increase, because water vapor (steam) has a lower density than liquid water. These conditions will remain the same until the last drop of liquid is vaporized. At this point, the entire cylinder is filled with vapor at 100 °C. This state is called **saturated vapor** (point 4 in Fig. 2.5). The state between saturated liquid (point 2 —

[c]A piston–cylinder apparatus is a standard hypothetical tool used to illustrate thermodynamic processes. The apparatus consists of a vertical container (the cylinder) of unlimited height with one mobile wall (the piston). The piston can move without friction inside the cylinder in response to the variation of the internal or external pressure. The weight of the piston will be considered negligible.

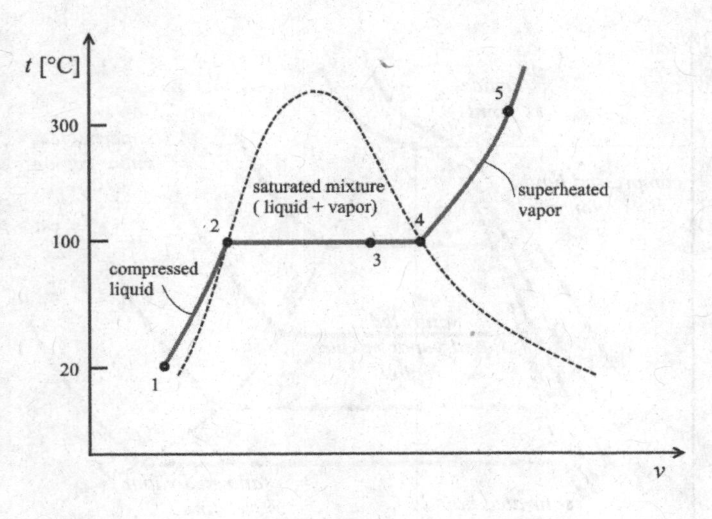

Fig. 2.5

liquid only) and saturated vapor (point 4 — vapor only) where two phases coexist at equilibrium is called **saturated liquid–vapor mixture** (point 3 in Fig. 2.5). Past point 4, the saturated vapor state, any addition of heat will increase the temperature and specific volume of the vapor. This state is called **superheated vapor** (point 5 in Fig. 2.5).

The heating process described above can be repeated at various pressures above and below 1 atm. That will result in a family of curves similar to 12345.

The locus of saturated liquid points (2) is a curve known as the **saturated liquid line**, while the locus of saturated vapor points (4) is a curve called **saturated vapor line** (Fig. 2.6). The symbols for state properties of saturated liquid take the subscript "f" (e.g., v_f, u_f, etc.). The symbols for state properties of saturated vapor take the subscript "g" (e.g., v_g, u_g, etc.). Notice that as the applied pressure is increased, the region between the saturated liquid and saturated vapor lines decreases until the two curves meet at the critical point [7, p. 60], where phase boundaries vanish.

This concept can be also applied to any other pure substance.

For the region enclosed between the saturated liquid line and the saturated vapor line (also referred to as the *saturated liquid–vapor*

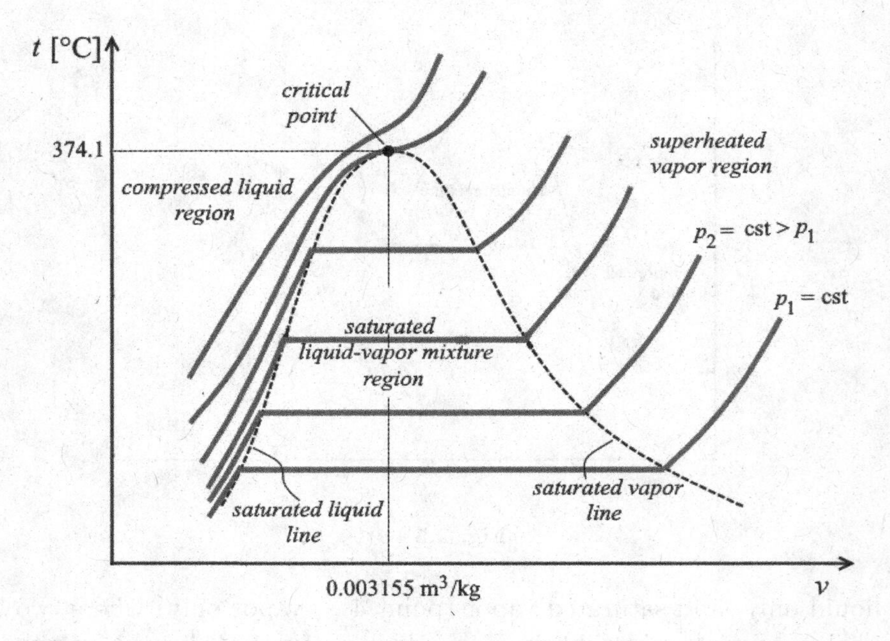

Fig. 2.6

mixture region), a factor x called the **quality** of the mixture (also referred to as the *dryness factor*) is defined as the mass of vapor divided by the total mass of the fluid [7, p. 63].

$$x = \frac{m_{\text{vapor}}}{m_{\text{liquid}} + m_{\text{vapor}}} = \frac{m_g}{m_f + m_g} = \frac{m_g}{m} \tag{2.1}$$

The value of the quality ranges from zero (at saturated liquid state) to 1 (at saturated vapor state). Sometimes the quality is given as a percentage. Outside the saturated liquid–vapor mixture region, the quality factor has no meaning.

In order to evaluate the quality x, let us consider a volume V containing a mass m of saturated liquid–vapor mixture (see Fig. 2.7 [7, p. 67]).

$$\begin{aligned} V &= V_f + V_g \\ mv &= m_f v_f + m_g v_g \\ mv &= (m - m_g)\, v_f + m_g v_g \end{aligned} \tag{2.2}$$

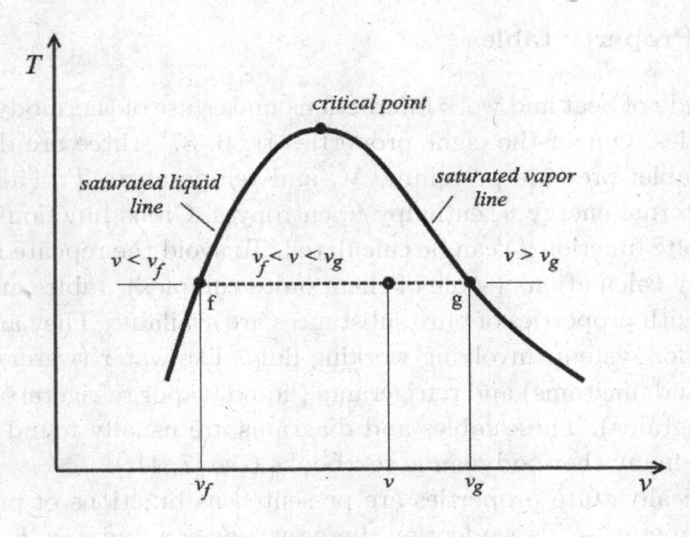

Fig. 2.7

According to Eq. (2.1)

$$x = \frac{m_g}{m} \quad \text{or} \quad m_g = x\,m$$

Dividing (2.2) by m one obtains

$$v = (1 - x)v_f + xv_g \tag{2.3}$$

$$v = v_f + xv_{fg} \tag{2.4}$$

or

$$x = \frac{v - v_f}{v_{fg}} \tag{2.5}$$

where

$$v_{fg} = v_g - v_f$$

The values v_g and v_f can be found in property tables (see Sec. 2.3). Conversely, Eqs. (2.3)–(2.5) can be used to evaluate the specific volume of the mixture v if x, v_g and v_f are known. These equations are valid for any other thermodynamic property like specific internal energy u, enthalpy h, and entropy s which will be defined in future sections.

2.3. Property tables

The study of heat and work interactions makes use of thermodynamic properties. Out of the eight properties [1, p. 87], three are directly measurable: pressure p, volume V, and temperature T. The other five (internal energy u, enthalpy h, entropy s, Gibbs function g, and Helmholtz function f) can be calculated. To avoid the repeated use of property calculations (some of them quite complex), tables and diagrams with properties of pure substances are available. They are used mainly for systems involving working fluids like water (water/steam tables and diagrams) and refrigerants (liquid/vapor refrigerant tables and diagrams). These tables and diagrams are usually found at the back of many thermodynamic textbooks (see [7, 11]).

Typically state properties are presented as functions of pressure and temperature; dependent on these two one can find v, u, h, and s.

For instance, in Appendix E one can find:

- Water properties (in SI units)
  - Saturated water/steam: Table A.2 (as function of t_{sat}); Table A.3 (as function of p_{sat})
  - Superheated steam: Table A.4
- Properties of saturated refrigerant R134a: Table A.6

In [7, Table T-5] one can find properties of compressed liquid water. There are also other tables and charts available for other thermophysical properties of gases, liquids and solids; e.g., properties of air in [7, Table T-9], etc.

When thermodynamic properties are presented in tabular form, data are indicated only at discrete points. For example, the value of the specific internal energy u for some gas (at some constant pressure) is indicated only for given temperatures: $u_1(t_1)$, $u_2(t_2)$, $u_3(t_3)$, etc. To estimate the value of u at any value of t other than t_1, t_2, t_3, etc. interpolation between adjacent table entries must be used. In most textbooks data is presented such that *linear interpolation* gives acceptable accuracy. An example of the use of linear interpolation can be found in [7, p. 66].

It must be noted that the absolute values of some calculated properties (like internal energy u and enthalpy h of a substance) cannot be determined as it is not possible to determine the exact values for some constituent energies. This is not an impediment, though, because it is the *change* in the value of a property accompanying a thermodynamic process that is of interest, and that can be calculated. If ΔX means a finite change (a difference) in the value of property X, then

$$\Delta X = X_{\text{final.state}} - X_{\text{initial.state}} \tag{2.6}$$

A negative value of ΔX indicates that the value of X has decreased during a process. Conversely, a positive value of ΔX indicates an increase in the value of X.

Because the absolute values of u and h are irrelevant, one can choose arbitrarily a **reference state** to which these properties can be assigned some **reference value** [7, p. 72]. For example, the reference state in the steam tables is the triple point of water ($t = 0.01$ °C, $p = 0.6113$ kPa) where the internal energy and entropy of saturated liquid water are assigned a value of zero (see [7, 11]); the value of the specific enthalpy h is calculated using $h = u + pv$ (see Chapter 3).

For refrigerants, one convention (NIST/ASHRAE [15]) is to use −40 °C (−40 °F) as a reference point for enthalpy and entropy, where h and s are considered zero for the saturated liquid; the same reference point is used in [7]. Some other sources [16] use 0 °C as a reference point, where $h_f = 200$ kJ/kg and $s_f = 1$ kJ/(kg K). This is also the reference point considered in Appendix E. Any convention is valid as long as sources are not mixed during calculations.

2.4. The ideal gas equation of state

Let us consider some gas reaching a large number of widely different equilibrium states for which pressure p, specific volume v and absolute temperature T will be measured. If for each of these states the quantity pv/T is calculated and plotted against p having T as a parameter, a family of T = cst curves is obtained, as depicted in Fig. 2.8 [1]. As pressure approaches zero, all these curves will converge

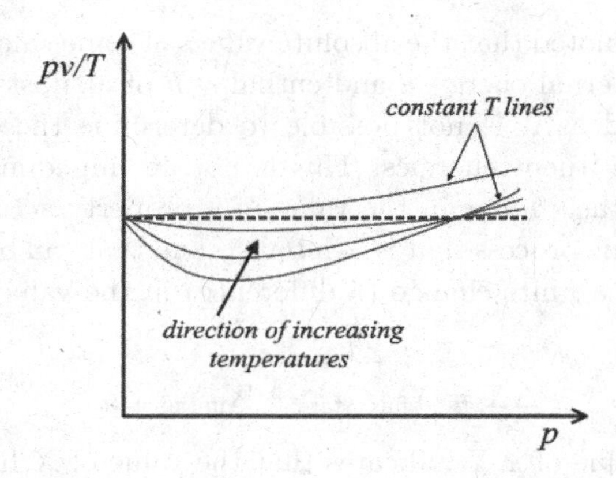

Fig. 2.8

to the same point. When repeating the procedure with other gases, the same type of graph will be obtained. Therefore it is be possible to write, for any gas,

$$\lim_{p \to 0} \left(\frac{pv}{T} \right) = A$$

where A is a constant, different for each gas.

Although the curves in Fig. 2.8 suggest that the mathematical relationship linking p, v and T is quite complex, for practical purposes, under ordinary conditions, the equation of state can be expressed quite accurately [1] by

$$pv = AT$$

If this equation were perfectly valid for any pressure and temperature combination, the representative curve would be the horizontal dashed line in Fig. 2.8. A hypothetical substance that satisfies the above equation is called a **perfect gas** or an **ideal gas**.[d] Under these circumstances

$$\frac{pv}{T} = A = \text{cst}$$

and the constant A is referred to as the **specific gas constant**, or simply the **gas constant**, denoted by symbol R. Therefore the

[d] A perfect gas obeys the ideal gas equation and has constant specific heats (see section 6.8). An ideal gas obeys the ideal gas equation and has temperature-dependent specific heats.

ideal gas equation of state becomes

$$pv = RT \tag{2.7}$$

Measuring units for R are: $[R]_{SI}$ = kJ/(kg K); $[R]_{\text{US Customary}}$ = Btu/(lb °R), or ft lbf/(lb °R).

For a mass m of ideal gas,

$$pV = mRT \tag{2.8}$$

In Eqs. (2.7) and (2.8) T must be thermodynamic temperature [K, or °R] and p — absolute (not gauge) pressure.

The approach on the ideal gas equation of state presented above assumes that the thermodynamic temperature T can be measured independently, using a gas thermometer [1, p. 157].

Historically speaking, the ideal gas equation of state was determined experimentally based on Boyle's law, Charles' law, and Avogadro's law. The result was

$$pV = n\bar{R}T \tag{2.9}$$

where n is the number of moles of substance and $\bar{R}$ is a constant known as the **universal gas constant**. Its value, as determined experimentally is [7]

$$\bar{R} = 8.314 \text{ kJ/(kmol K)} = 1.986 \text{ Btu/(lbmol °R)}$$

and is the same for all gases. When dividing equation (2.9) by n, one obtains

$$p\frac{V}{n} = \bar{R}T$$

$$p\bar{v} = \bar{R}T \tag{2.10}$$

where $\bar{v} = \frac{V}{n}$ is the molar volume = 22.41 m^3/kmol (at 273.15 K and 101.325 kPa) [17].

When dividing and multiplying the right side of Eq. (2.9) by M, the molar mass,

$$pV = nM\frac{\bar{R}}{M}T$$

one obtains

$$pV = mRT$$

where $nM = m$ and $\bar{R}/M = R$.

Table 2.1

Property	**Ideal gas**	**Real gas**
Obeys the equation $pv = RT$	always	only at low pressures and high temperatures
Identical molecules	always	only for pure gases
Molecular volume	zero	small, but not zero
Intermolecular forces	zero	small, but not zero
Collisions	perfectly elastic	not perfectly elastic

The gas constant R is, therefore, a constant for the particular gas with molecular weight M.

With the improvement of the precision of measuring equipment, it has been determined that the equations of state (2.8) or (2.9) are not accurate for any state condition. They are strictly true only for "ideal gases". The concept of an ideal gas is associated to the simplest assumptions about its molecules, as presented in Table 2.1. **Real gases**[e] have a more complex behavior.

It is estimated [18], however, that over the normal working range of pressure and temperature, the ideal gas equation of state predicts the properties of real gases with an error of maximum 5 %. At a temperature higher than the critical temperature and at low pressures, the ideal gas law is a very good model for the gaseous phase; gas molecules behave like billiard balls: they are far enough apart for intermolecular forces to be negligible. The ideal gas model can be used when solving problems involving simple thermodynamic processes regardless of whether the gas has a single component (like oxygen, nitrogen, carbon dioxide and other common gases [7]) or is a mixture (like air).

At low temperatures and high pressures, the simplified model of the ideal gas no longer applies. Under these conditions, corrections have to be made to Eq. (2.7) by introducing terms to describe attractions and repulsions between molecules in order to better predict the properties. Therefore real gases require more complex equations of state than ideal gases. The van der Waals equation

[e]See also [67].

(1873) [1, pp. 106, 167] is one of the common corrections made to the ideal gas law, but there are also other methods dealing with real gases: Clausius equation [19], Redlich–Kwong equation (1949) [20], virial equations [21], etc.

Properties of pure substances — problems

2.1. What happens to the saturation temperature if, during the boiling process, the pressure of the liquid changes?

2.2. What is the quality of steam, x, if for each kilogram of vapor at saturation we have 0.15 kg of saturated liquid?

2.3. A metallic air storage tank with a capacity of 10 L is filled, through a valve, with 2 kg of air considered a perfect gas having $R = 287$ J/(kg K). The valve is then closed. The air inside the container and the container itself are in thermal equilibrium with the surroundings, at $t = 24$ °C. Determine:

(a) what kind of thermodynamic system represents the gas in the container;
(b) the specific volume of the air in the container;
(c) the absolute pressure, p, of the air in the container.

2.4. A rigid container is filled with carbon monoxide ($R = 297$ J/(kg K)) through an inlet valve for 6 s at a constant mass flow rate of $\dot{m} = 0.3$ kg/min. The volume of the container is 9 L. After the inlet valve is closed and the gas reaches thermal equilibrium, the inner pressure reaches $p = 4$ kgf/cm^2. Determine the temperature of the gas at this point.

2.5. A rubber balloon is filled with helium ($R = 2077$ J/(kg K)) at temperature $t = 18$ °C. Determine: (a) the type of thermodynamic system characterizing the gas inside the balloon; (b) the specific volume of helium if the pressure inside the balloon is 2.4 bar.

Chapter 3

The First Law of Thermodynamics

There are two distinct *forms of energy*:

- *static* forms, that can be stored in the system and characterize the state of the system, and
- *dynamic* forms, that cannot be stored in the system, called interactions, that are recognized at the system boundary which they cross as the system changes its state.

Energy can cross the boundaries of a closed system, in the process of transfer from one body to another, in two distinct forms: **heat** and **work**. These are transient quantities; they exist as long as the system undergoes a change, therefore they cannot be associated to a state of the system.

3.1. Heat

Heat is the form of transfer of energy between two systems (or a system and its surroundings) which appears at the boundary of a system that changes its state due to a *temperature difference* between the two systems (or the system and its surroundings) [11]. Note that temperature difference is an indispensable condition for heat transfer. As a consequence, no heat transfer will occur between two systems that are at the same temperature.

In SI, heat is measured in joule [J] and its multiples. When associated to a thermodynamic process between states 1 and 2, symbol Q_{12} is used or, if there is no possibility of confusion, simply Q.

The amount of heat transferred per unit mass by a system of mass m can be expressed as

$$q = \frac{Q}{m} \quad [\text{J/kg}]$$

The rate of heat transfer can be expressed as

$$\dot{Q} = \frac{Q}{\tau} \quad [\text{W}]$$

where τ is the symbol for time. Notice the usual notation using an overdot to express the change in some quantity with respect to time (time derivative).

3.2. Work

Work is the form of transfer of energy which appears at the boundary of the system due to the *movement* of a part of the boundary under the action of a *force* [1]. Note that both these elements — displacement and force causing it — must exist in order to consider a transfer of energy in form of work.

In SI, work, too, is measured in joule [J] and its multiples. When associated to a thermodynamic process between states 1 and 2, symbol W_{12} is used or, if there is no possibility of confusion, simply W.

The amount of work transferred per unit mass by a system of mass m can be expressed as

$$w = \frac{W}{m} \quad [\text{J/kg}]$$

The rate of energy transfer in form of work can be expressed as

$$\dot{W} = \frac{W}{\tau} = P \quad [\text{W}]$$

where τ is the symbol for time and $\dot{W}$ represents power (P).

Heat and work are scalars; they are completely described by one single characteristic: magnitude. However, in thermodynamics we make a distinction between energy entering and energy leaving a system. Therefore some sign convention is necessary for heat and work. In this book, we will use the following convention[a]:

- when heat is transferred *into* a system *from* the surroundings, the quantity is considered *positive.* Conversely, when heat is transferred *from* a system *to* the surroundings, the quantity is considered *negative*;
- when work is done *by* the system *on* the surroundings (e.g., when a fluid expands pushing a piston outwards), the work is said to be positive. Conversely, when work is done *on* the system *by* the surrounding (e.g., when the fluid is compressed), the work is said to be *negative.*

Heat and work are transient quantities. They only appear at the boundary while a change is taking place within the system. They cannot be stored in the system and therefore cannot define the state of the system; they are not properties (state functions) but **path functions**. Path functions do not have exact differentials. According to a convention generally accepted in thermodynamics, symbol d is reserved for exact total differentials, that is for properties or state functions (like p, V, T). For infinitely small quantities that do not refer to properties, the symbol δ is used (e.g., δW or δQ). Also, during some finite process, the symbol Δ is reserved for small variations of state functions, while finite amounts of heat or work are simply denoted by Q or W.

For a process between states 1 and 2, for a state function (like V) one can write[b]

$$\int_1^2 dV = V_2 - V_1 = \Delta V, \quad \text{but} \int_1^2 \delta W = W_{12} \quad (\text{not } \Delta W).$$

[a]The sign convention is arbitrary. The convention described above (adopted in [7] and [11]) dates probably from the time of the first thermal machines (steam engines): heat transferred into a system and work output from a system are both taken as positive. A different convention is adopted in [1]: energy entering the system is positive, energy leaving is negative.

[b]State functions can be treated mathematically as vector functions [68]. For conservative vector fields, the line integral is independent of path and is equal to the difference between the values of the potential function at path ends.

In many thermodynamic problems, mechanical work is the only form of work involved. This requires: (*a*) a *force* acting on the boundary, and (*b*) a *displacement* of (a part of) the boundary.

3.3. The cycle

A *closed system* is said to undergo a cyclic process, or **cycle**, when it passes through a series of states in such a way that its final state is equal in all respects to its initial state [1, p. 15]. Since, at the end of the cycle, all properties have regained their initial values, the system is in the position to be put through the same cycle again, and the procedure may be repeated indefinitely. This is the reason why a cyclic process is the underlying principle for any power and refrigeration plant. In a thermodynamic cycle, the overall change in a state function is zero.

The cycle consists then of a series of thermodynamic processes during which the working fluid performs energy exchanges (heat and work) with the surroundings. Heat and work are path functions, therefore they do not have an exact total differential and their cyclic integral is not zero.

A reversible cycle (*ideal cycle*) consists only of *reversible processes.* These processes can be graphically represented in a thermodynamic diagram ($p - V$, $T - s$, $h - s$, etc.).

3.4. The first law of thermodynamics

J. P. Joule carried out, after 1843, a series of experiments which indicated the relationship between heat and work in a thermodynamic cycle for a system. The description of his classic paddle wheel experiment can be found in any elementary textbook on physics. He used a rotating paddle wheel to stir the water from an insulated vessel, by lowering a weight of mass m by a distance Δz. He noted the amount of work done on the paddle wheel ($W = mg\Delta z$) and the increase in water temperature due to the friction between the water and the paddles. Later, the vessel was placed in a bath and cooled down, thus releasing heat Q estimated in terms of temperature increase of bath. Joule repeated the experiment for various amounts and rates of work done and concluded that the heat Q rejected from the system was proportional to the net work W supplied by the lowered weight. He established that heat is also a form of energy, like work. Before Joule, according to the *caloric theory*, heat was considered to be an invisible fluid flowing from a higher calorie to a lower calorie body [22, p. 71].

Joule's experiments, and others of the same nature, lead to the foundation of the **first law of thermodynamics** (simply called *first law*). Usually, the first law is stated this way: "*for a closed system, during a cycle, the net work delivered to the surroundings is proportional to the net heat taken from the surroundings*" [1, p. 17]. Mathematically, the first law can be written as

$$\left(\sum \delta Q\right)_{\text{cycle}} \propto \left(\sum \delta W\right)_{\text{cycle}}$$

or

$$\oint \delta Q \propto \oint \delta W \tag{3.1}$$

where the integral symbol with a centered circle $\oint$ denotes a *circular integral*, or *cycle integral* for the closed path (see also Sec. 6.2). The proportionality can be turned into equality by introducing a constant of proportionality:

$$J \oint \delta Q = \oint \delta W \tag{3.2}$$

where the constant J is the so-called *mechanical equivalent of heat* or *Joule's equivalent.* Since heat and work are both forms of energy transfer, if appropriate measuring units are used so that the constant of proportionality J becomes 1, then one can write that

$$\oint \delta Q = \oint \delta W$$

or

$$\oint \delta Q - \oint \delta W = 0 \tag{3.3}$$

Here the "–" sign in Eq. (3.3) belongs to the formula; $\oint \delta Q$ represents the *net heat*, which can be regarded as an algebraic sum of positive and negative quantities of heat (according to the sign convention), while $\oint \delta W$ represents the *net work* which can also be regarded as an algebraic sum of positive and negative quantities of work associated to the cycle.

It should be noted that the first law of thermodynamics is similar to the **conservation of energy principle**, according to which energy cannot be created nor destroyed during a process; it can only change forms.

3.5. Corollaries of the first law of thermodynamics

Corollary 1. *There exists a property of a closed system such that a change in its value is equal to the difference between the heat supplied and the work done during any change of state* [1].

Proof. For this, the technique known as "reductio ad absurdum" (reduction to absurdity) will be used; i.e., assume that the opposite of the proposition is true, namely that $\int_1^2 (\delta Q - \delta W)$ depends upon the process, and show this to be an absurdity because it contradicts the first law.

Consider a system that changes its state from 1 to 2 by some processes A and B (Fig. 3.1). Let us assume that in general

$$\int_1^2 (\delta Q - \delta W)_A \neq \int_1^2 (\delta Q - \delta W)_B \tag{3.4}$$

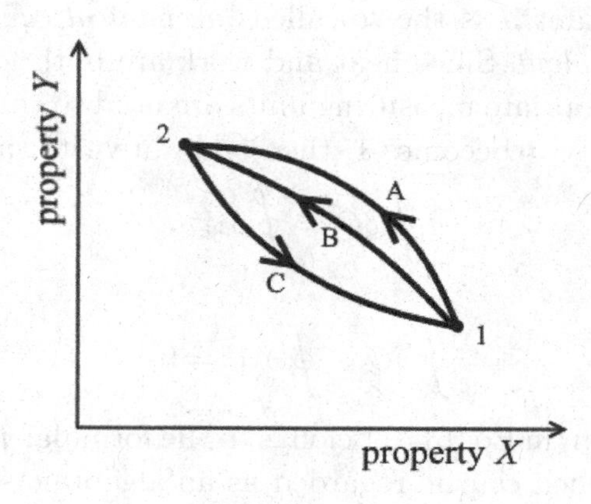

Fig. 3.1

where indices A and B show the process for which the summation is performed. Now let also assume that in each case the system returns to its original state by a process C. Two cycles are formed: $1A2C1$ and $1B2C1$. Then, we have

$$\oint_{1A2C1} (\delta Q - \delta W) = \int_1^2 (\delta Q - \delta W)_A + \int_2^1 (\delta Q - \delta W)_C$$

and

$$\oint_{1B2C1} (\delta Q - \delta W) = \int_1^2 (\delta Q - \delta W)_B + \int_2^1 (\delta Q - \delta W)_C$$

If the assumption described by inequality (3.4) is true, it follows that

$$\oint_{1A2C1} (\delta Q - \delta W) \neq \oint_{1B2C1} (\delta Q - \delta W)$$

which is absurd because it contradicts the first law expressed by Eq. (3.3); the two cycle integrals should be equal since they are both zero. Therefore the original proposition, that $\int_1^2 (\delta Q - \delta W)$ is independent of the process, must be true.

If the newly discovered property is denoted by U, the corollary can be expressed as

$$\int_1^2 (\delta Q - \delta W) = U_2 - U_1$$

or, after integration,

$$Q - W = \Delta U = U_2 - U_1 \tag{3.5a}$$

For one unit of mass of substance (1 kg):

$$q - w = \Delta u = u_2 - u_1 \tag{3.5b}$$

The property U is called the **internal energy** of the system and Eq. (3.5a) is called the **non-flow energy equation**. Again, U, the energy of a system due to its temperature,[c] is a property (state function), unlike Q and W. A loss of internal energy implies a temperature drop.

[c]For ideal gases, the internal energy depends only on temperature [69]. See also Joule and Gay-Lussac's experiment with free expansion described in [70]. For real gases, the internal energy depends on temperature and pressure.

In SI, internal energy is measured in joule [J] and its multiples. Now $u = U/m$ (where m is the mass of the system) is the *specific internal energy*, just as v represents the specific volume.

The equation of definition of internal energy suggests no means for assigning an absolute value to the internal energy of a system. This is why, for practical purposes, an arbitrary reference state was considered (see Sec. 2.3) for which a reference value U_0 is attributed. The values found in thermodynamic tables for various states, U_1, U_2, U_3, etc. are, strictly speaking, $(U_1 - U_0)$, $(U_2 - U_0)$, $(U_3 - U_0)$, etc. When analyzing a process between states 1 and 2, the associated change in internal energy $(U_2 - U_1)$ is actually calculated as $(U_2 - U_0) - (U_1 - U_0)$, where U_0 cancels. This is why in all engineering applications of thermodynamics where only changes in energy levels are important, the reference state/value is irrelevant.

In the analysis of certain types of processes, a frequently encountered combination is $(U + pV)$. This is defined as a new property called **enthalpy**, with the symbol H:

$$H = U + pV \tag{3.6a}$$

In SI, enthalpy is measured in Joule [J] and its multiples. The *specific enthalpy* of a system of mass m is as follows:

$$\frac{H}{m} = h = u + pv \quad \text{[J/kg]} \tag{3.6b}$$

For an ideal gas, Eq. (3.6b) can be written as $h = u + RT$ which suggests that for an ideal gas enthalpy depends only on temperature.

Corollary 2. *The internal energy of a closed system remains the same if the system is isolated from its surroundings.*

No formal proof is required for this statement once the first corollary has been proved, because it is a direct consequence of Eq. (3.5a). For an isolated system the exchange of energy with the surroundings being zero, it follows that Q and W are zero. Therefore ΔU must be zero.

Corollary 2 is often called the *law of conservation of energy* [1, p. 22].

Corollary 3. *A perpetual motion machine of the first kind is impossible.*

Once started, such a machine would continuously supply work without absorbing energy from the surroundings. In Eq. (3.3) if the heat integral is zero, the work integral is also zero. Therefore no machine can work in a cycle without any energy input. Certainly, it is always possible to obtain a limited amount of work from a thermodynamic process without the presence of an external source. For instance, if a gas is compressed in a piston–cylinder apparatus, once the piston is released the gas will expand producing some work without any external heat input. However, the process will stop once the state of equilibrium with the surroundings is reached and no cycle will be possible.

3.6. Energy balances

The *conservation of energy principle* can be expressed as follows: "the net change (increase or decrease) in the total energy of the system during a process is equal to the difference between the total energy entering the system and the total energy leaving the system during that process" [11]:

$$E_{\text{in}} - E_{\text{out}} = \Delta E_{\text{system}} \tag{3.7}$$

This relation is referred to as **the energy balance**. Also, any finite change in the energy level of a system is as follows:

$$\Delta E_{\text{system}} = E_{\text{final}} - E_{\text{initial}} = E_2 - E_1 \tag{3.8}$$

3.6.1. *Energy balance for closed systems*

For a closed system, energy can enter or leave the system only through heat transfer (Q) and work (W). Therefore

$$E_{\text{in}} - E_{\text{out}} = Q - W = (Q_{\text{in}} - Q_{\text{out}}) - (-W_{\text{in}} + W_{\text{out}})$$

or

$$E_{\text{in}} - E_{\text{out}} = (Q_{\text{in}} - Q_{\text{out}}) + (W_{\text{in}} - W_{\text{out}}) = \Delta E_{\text{system}} \tag{3.9}$$

3.6.2. *Energy balance for open systems*

Energy balances on open systems have to take into account the energy associated with the mass that flows through the system. Therefore

$$E_{\text{in}} - E_{\text{out}} = (Q_{\text{in}} - Q_{\text{out}}) + (W_{\text{in}} - W_{\text{out}}) + (E_{\text{mass,in}} - E_{\text{mass,out}})$$
$$= \Delta E_{\text{system}} \tag{3.10}$$

In a rate form [11]:

$$\dot{E}_{\text{in}} - \dot{E}_{\text{out}} = \frac{dE_{\text{system}}}{d\tau}$$

where τ represents time.

For one unit of mass,

$$e_{\text{in}} - e_{\text{out}} = \Delta e_{\text{system}}$$

Energy balances are used in many applications involving power plants and other industrial units to identify losses.

3.7. Effectiveness of energy conversion

Engineering thermodynamics studies essentially phenomena related to conversion of energy from one form to another. A power plant, for example, transforms thermal energy, often obtained from combustion of a fuel, into mechanical energy and then into electrical energy. In an internal combustion engine powering a vehicle, the chemical energy of fuel is converted into mechanical energy of the vehicle. A measure of the effectiveness of the energy conversion process is what we will call the *performance index*. This will be defined as

$$\text{Performance index} = \frac{\text{Desired output}}{\text{Required input}}$$

By extension, the performance index, qualitatively described by the above relationship, can be used to characterize any thermodynamic cycle. Depending on the case, the performance index is referred to as **efficiency** (for power cycles) or **coefficient of performance** (for refrigeration cycles).

The first law of thermodynamics — problems

3.1. A mass of air undergoes a thermodynamic process such that it does 25 kJ of work while it receives 18 kJ of heat. Another process is conducted such that, while expanding between the same thermodynamic states, the system requires a heat input of only 10 kJ. Determine the change of internal energy during the first process and the work done by the air in the second expansion.

3.2. Determine the enthalpy of 1 m^3 of air (ideal gas) at 177 °C ($u = 323$ kJ/kg).

3.3. A thermodynamic system is said to produce 915 kJ of work during a process while receiving only 630 kJ of heat. Is it possible? What is the corresponding change of internal energy? What if, during the same heat input, the system receives the amount of work from the surroundings?

3.4. Heat and work are: (a) transient forms of energy; (b) internal energies; (c) intrinsic forms of energy; (d) extrinsic energies.

3.5. The sum of internal energy u and product pv is called: (a) specific energy; (b) specific enthalpy; (c) specific entropy; (d) total energy.

Chapter 4

Non-flow Thermodynamic Processes of an Ideal Gas

In this chapter, we will deal with non-flow energy equations expressing the first law written for various reversible processes occurring in closed systems.

4.1. The energy equation and reversibility

In Sec. 3.4 the non-flow energy equation was presented in the form (3.5a)

$$Q - W = \Delta U = U_2 - U_1$$

or, for 1 kg of mass,

$$q - w = u_2 - u_1 \tag{4.1}$$

For what we call **reversible processes** [7, p. 128], the thermodynamic system passes through a continuous series of equilibrium states. This condition is satisfied only if the process is due to infinitesimal changes in some property of the system. Consequently, it would take an infinite amount of time to complete a reversible process. This makes a perfectly reversible process impractical. However, if the relaxation time of the system is much shorter than the duration of the process, the deviation from reversibility may be negligible (quasi-static process).[a] In such cases, Eq. (4.1) may be applied to

[a]See Sec. 1.4.

any *infinitesimal part* of the process between the end states:

$$\delta q - \delta w = du$$

For a reversible process, the overall changes occurring between the end states can be found by integration:

$$q_{12} - w_{12} = \int_1^2 du = u_2 - u_1$$

During **irreversible processes**, the system does not pass through a series of equilibrium states. It is not, therefore, possible to write $U = mu$ since (a) a single value of u cannot be used to define the random molecular energy of the fluid (each elemental mass has its own specific internal energy), and (b) this energy is not the only one relevant form if the fluid is turbulent (there are also the kinetic and the potential energy of the mass). Consequently, an irreversible process cannot be represented in a thermodynamic diagram with a continuous line (a series of equilibrium states). Also, the energy equation can be used only in the integrated form (3.5a).

For a system to undergo a reversible process, work and heat must cross the boundaries of a system under certain condition, as will be discussed in the next sections.

4.1.1. *Work and reversibility*

Let us consider a closed system (Fig. 4.1) constituted by some gas inside a piston–cylinder apparatus. Part of the boundary (adjacent to

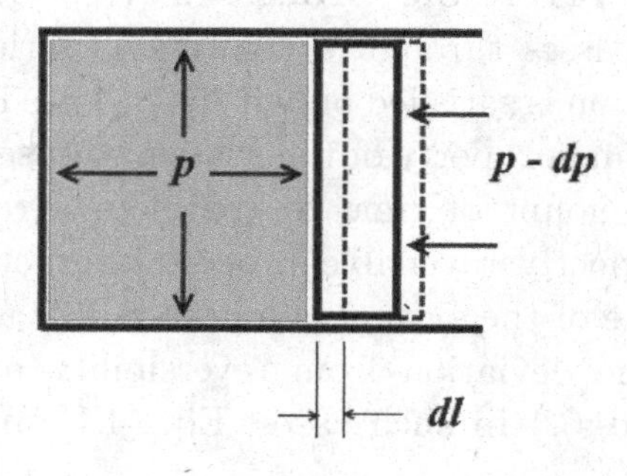

Fig. 4.1

the piston) is allowed to move under such conditions that the external pressure is always only infinitesimally smaller than the pressure of the system p. This pressure is assumed constant during an infinitesimal movement of the piston over a distance dl. If A is the area of the piston, the force created by pressure p is $F = pA$.

The infinitesimal amount of work produced by the system is

$$\delta W = F\,dl = pA\,dl = p\,dV$$

If the process occurs from p_1 to p_2 in such a way that the restraining force is changing continuously so that it is never significantly different from pA, the total work can be found by summing up all the increments of work $p\,dV$ [7, p. 36]:

$$W = \int_1^2 p\,dV \qquad (4.2)$$

The conditions under which this expansion has been considered are the ones required for reversibility:

- no pressure gradients or eddies are set up in the fluid;
- properties are uniform throughout the system at all times.

The process can be represented by a full line on the $p - V$ diagram (Fig. 4.2). The area under this curve is given by $\int_1^2 p\,dV$ which represents, by comparison to (4.2), the work done by the fluid. Since the process is *reversible*, the expansion may be stopped at any point and reversed by only an infinitesimal change in the external force. The system will return through the same series of states, and exactly the same amount of work will be done by the surroundings on the system [7, p. 37].

An example of irreversible process is the *free expansion*: an insulated container is divided in two compartments by a removable partition. One compartment contains some gas at state (p_1, v_1) while the other one is evacuated. When the partition is removed, the gas expands freely because the process is not restrained in any way ($W = 0$). Once the system reaches equilibrium, its new state is (p_2, v_2). The end states can be represented on a p–v diagram, but not the intermediate states, because they are not states of equilibrium [1, p. 22].

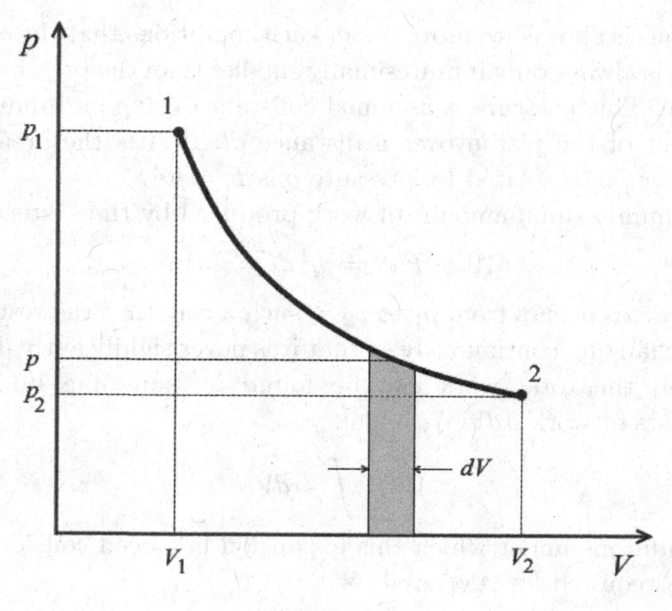

Fig. 4.2

4.1.2. *Heat and reversibility*

The heat exchange between a system and its surroundings makes use of the concept of **heat reservoir**; this is a part of the surroundings which exchanges energy with a system because it is at a different temperature. A heat reservoir may be a *source* or a *sink* of heat [1, p. 27].

A heat reservoir is considered to have *infinite capacity*, such that any amount of heat exchanged during a process will not change its temperature. In industrial applications, for example, a source can be the continuous combustion of a fuel, or a mass of condensing vapor; a sink can be a river, the earth's atmosphere, or a mass of melting solid.

If we need to raise the temperature of a system consisting of a fluid in a closed container from 200 to 300 °C using a source at 400 °C, heat will cross the boundary at a finite rate. This will create a temperature gradient in the fluid and therefore the system cannot be in a state of equilibrium during the transfer of energy. If the source is replaced by a series of heat sources, ranging in temperature from $200 + d\theta$ to $300 + d\theta$ so that the active source is always only infinitesimally higher in temperature than the system, temperature gradients will not be produced in the fluid and the system may be presumed to pass through a series of equilibrium states. At any point in the process only an infinitesimal change in the temperature of the source is necessary to turn it into a sink and reverse the process.

In conclusion, a reversible transfer of energy by virtue of temperature difference can only be carried out if the temperature difference is infinitesimally small [1, p. 28]. In this case, the rate of heat transfer is infinitely slow and therefore such a reversible process cannot be completely achieved in practice. However, there are practical methods to achieve reversible heat transfer (see Sec. 6.6).

For an irreversible non-flow process, the energy equation can be applied in the integrated form:

$$q - w = u_2 - u_1$$

because states 1 and 2 are states of equilibrium. For reversible processes, the differential form can also be used:

$$\delta q - \delta w = du$$

or

$$\delta q - p\,dv = du \tag{4.3}$$

In Sec. 4.1.1 it was shown that work w (a transient quantity) can be expressed in terms of two properties, p and v. By analogy, it might be supposed that heat too can be expressed in terms of properties when transferred during a reversible process. As temperature difference is the driving force for heat transfer, temperature might be one of the properties. For example, we might be able to write $\delta q = \theta\,dx$ where θ is the temperature and x some other property [1, p. 28]. Indeed, it will be shown in Chapter 6 that we can write $\delta q = T\,ds$ where T is the *thermodynamic* (*absolute*) *temperature* and s is a property called *entropy.*

4.2. Isochoric process

This is a process during which the volume (or specific volume) remains constant (*isos* Greek = equal; *choras* Greek = space). Sometimes it is called *isometric process.* Its characteristic equation is

$$V = \text{cst} \quad (\text{or } v = \text{cst})$$

From the equation of state,

$$pv = RT \Rightarrow \frac{pv}{T} = \text{cst}$$

If v = cst, then the relation between properties for the isochoric process is

$$\frac{p}{T} = \text{cst} \quad \text{or} \quad \frac{p_1}{T_1} = \frac{p_2}{T_2} \quad \text{or} \quad \text{even} \quad \frac{p_1}{p_2} = \frac{T_1}{T_2} \tag{4.4}$$

Since v = cst, then $dv = 0$ so, for one unit of mass (1 kg) of ideal gas,

$$\delta w = p\,dv = 0 \tag{4.5}$$

i.e., no work is involved. From Eq. (4.3), the heat exchanged is

$$\begin{aligned} \delta q &= du \\ q &= u_2 - u_1 \end{aligned} \tag{4.6}$$

The enthalpy change is

$$dh = d(u + pv) = du + d(pv) = du + v\,dp + p\,dv = du + v\,dp \tag{4.7}$$

Integrating (4.7) one obtains

$$\begin{aligned} \int_1^2 dh &= \int_1^2 (du + v\,dp) \\ h_2 - h_1 &= u_2 - u_1 + v(p_2 - p_1) \end{aligned} \tag{4.8}$$

Also, because $pv = RT$,

$$\begin{aligned} d(pv) &= R\,dT \\ v\,dp + p\,dv &= R\,dT \\ v\,dp &= R\,dT \end{aligned}$$

After integration,

$$v(p_2 - p_1) = R(T_2 - T_1) \tag{4.9}$$

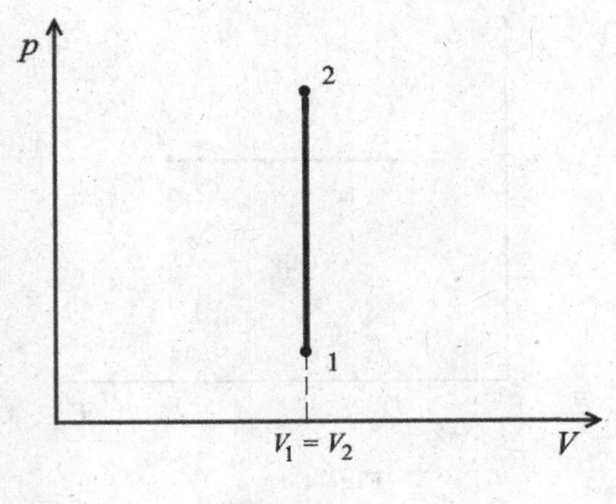

Fig. 4.3

The graphical representation of the isochoric process in the p–V diagram is a vertical line (Fig. 4.3). Since there is no area under the line, no work is done.

4.3. Isobaric process

This is a process during which the pressure remains constant (*baros* Greek = weight).

$$p = \text{cst}$$

From the equation of state,

$$\frac{V_1}{V_2} = \frac{T_1}{T_2} \tag{4.10}$$

The work associated with the process (for one unit of mass):

$$\begin{aligned} pv &= RT \\ d(pv) &= R\,dT \\ v\,dp + p\,dv &= R\,dT \quad (dp = 0) \\ p\,dv &= R\,dT \\ w = p(v_2 - v_1) = pv_2 - pv_1 &= R(T_2 - T_1) \end{aligned} \tag{4.11}$$

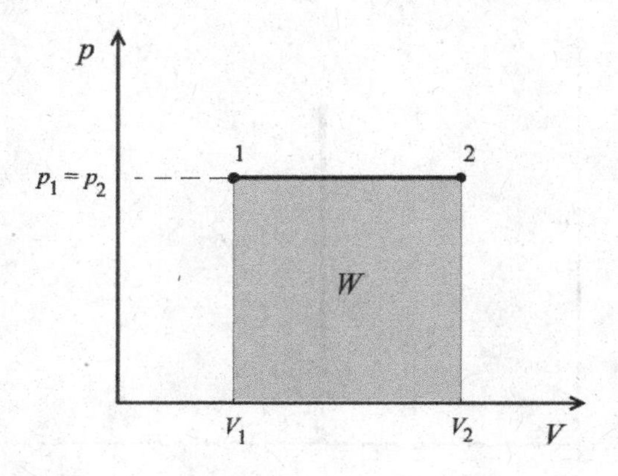

Fig. 4.4

From Eq. (4.3), the heat exchanged can be calculated:

$$\begin{aligned} \delta q - p\,dv &= du \\ q - p(v_2 - v_1) &= u_2 - u_1 \\ q &= u_2 - u_1 + pv_2 - pv_1 \\ &= (u_2 + pv_2) - (u_1 + pv_1) = h_2 - h_1 \end{aligned} \tag{4.12}$$

The graphical representation of the isobaric process in the p–V diagram is a horizontal line (Fig. 4.4). The amount of work is represented by the area under the curve.

4.4. Isothermal process

This is a process during which the temperature remains constant.

$$T = \text{cst}$$

From the equation of state,

$$pv = \text{cst}$$

$$\Rightarrow \frac{p_1}{p_2} = \frac{v_2}{v_1} \quad \text{or} \quad \frac{p_1}{p_2} = \frac{V_2}{V_1} \tag{4.13}$$

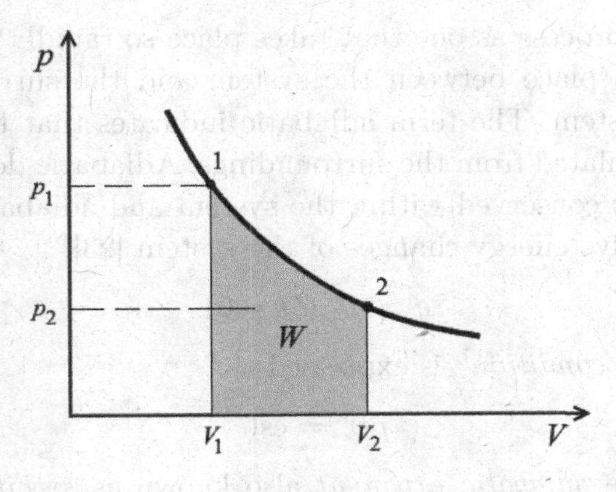

Fig. 4.5

The work associated with the process (for one unit of mass):

$$w = \int_1^2 p\,dv = \int_1^2 \frac{RT}{v} dv = RT \int_1^2 \frac{1}{v} dv = RT[\ln v]_1^2 = RT \ln \frac{v_2}{v_1} \tag{4.14}$$

For a given amount of ideal gas its internal energy is only a function of temperature. Therefore

$$T = \text{cst} \Rightarrow u = \text{cst}$$

Also, $h = \text{cst}$. The heat exchanged in the process is

$$\delta q - \delta w = du = 0$$

$$q = w = RT \ln \frac{v_2}{v_1} \tag{4.15}$$

The graphical representation of the isothermal process in the p–V diagram is a rectangular (or equilateral) hyperbola (Fig. 4.5); the asymptotes are the p-axis and the V-axis. The amount of work is represented by the area under the curve.

4.5. Adiabatic process

This is a process during which no heat crosses the boundary of the system (*diabainein* Greek = to cross, go through). One can think of an adiabatic process as one that takes place so rapidly that no heat transfer takes place between the system and the surroundings, or within the system. The term adiabatic indicates that the system is thermally insulated from the surroundings. Adiabatic does not mean that energy is conserved within the system and adiabatic processes typically involve energy changes of the system [23].

$$q = 0 \quad (Q = 0)$$

The *adiabatic condition*[b] is expressed as

$$p\,v^k = \text{cst} \tag{4.16}$$

where k is the *adiabatic exponent* also known as *specific heat ratio* (see Sec. 6.7). For simplicity, we will assume that for an ideal gas, k is a constant [24] depending on the nature of the gas; k is always greater than 1 (see also Sec. 6.7).

For an adiabatic process between states 1 and 2, $p_1 v_1^k = p_2 v_2^k$. This can be written as

$$p_1 v_1 v_1^{k-1} = p_2 v_2\, v_2^{k-1}$$
$$RT_1 v_1^{k-1} = RT_2 v_2^{k-1}$$

or

$$\frac{T_1}{T_2} = \left(\frac{v_2}{v_1}\right)^{k-1} \tag{4.17}$$

From the ideal gas equation of state:

$$\frac{p_1 v_1}{T_1} = \frac{p_2 v_2}{T_2} \Rightarrow \frac{p_1^k v_1^k}{T_1^k} = \frac{p_2^k v_2^k}{T_2^k}$$

or

$$\frac{p_1 p_1^{k-1} v_1^k}{T_1^k} = \frac{p_2 p_2^{k-1} v_2^k}{T_2^k}.$$

Since $p_1 v_1^k = p_2 v_2^k$, then

$$\frac{p_1^{k-1}}{T_1^k} = \frac{p_2^{k-1}}{T_2^k}$$

[b]The derivation of the adiabatic condition can be found in Appendix A.

or

$$\left(\frac{p_1}{p_2}\right)^{k-1} = \left(\frac{T_1}{T_2}\right)^k$$

$$\frac{T_1}{T_2} = \left(\frac{p_1}{p_2}\right)^{\frac{k-1}{k}} \tag{4.18}$$

In general, considering (4.17) and (4.18),

$$\frac{T_1}{T_2} = \left(\frac{v_2}{v_1}\right)^{k-1} = \left(\frac{p_1}{p_2}\right)^{\frac{k-1}{k}} \tag{4.19}$$

The work associated with the adiabatic process (for one unit of mass) is

$$\begin{aligned} \delta q - \delta w &= du \\ -\delta w &= du \\ w &= u_1 - u_2 \end{aligned} \tag{4.20}$$

The graphical representation of the adiabatic process in the p–V diagram has also a hyperbolic shape (Fig. 4.6). Like isotherms, every adiabat asymptotically approaches both the V-axis and the p-axis. An adiabat looks similar to an isotherm, except that an adiabat has

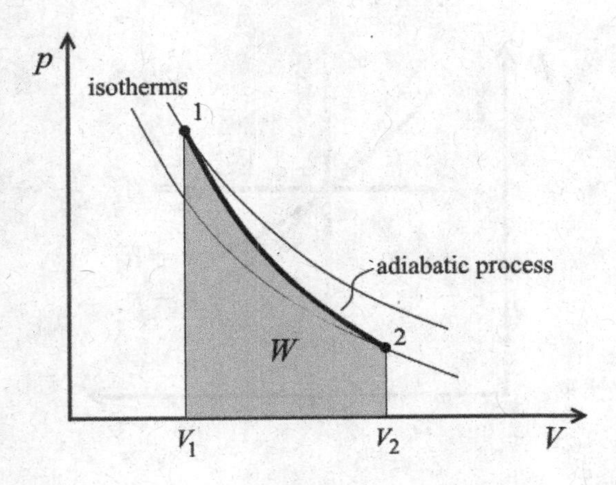

Fig. 4.6

a steeper inclination (closer to the vertical). Each adiabat intersects each isotherm only once.

4.6. Polytropic process

This is a process where p, V, and T change, while there is an exchange of energy in the form of heat and work with the surroundings. The relationship between the pressure and volume during a polytropic process of an ideal gas can be described analytically by the equation:

$$p\,v^n = \text{cst} \tag{4.21}$$

where n is a constant for the particular process. It is called *polytropic exponent* and is a real number.

For a process between states 1 and 2,

$$p_1 v_1^n = p_2 v_2^n = p v^n$$

Many processes can be approximated by the polytropic law (see Fig. 4.7):

- $n = 0$, results in $p = \text{cst}$, i.e., an isobaric process;
- $n = \infty$, results in $v = \text{cst}$, i.e., an isochoric process because

$$p^{1/n} v = \text{cst} \Rightarrow v = \text{cst}$$

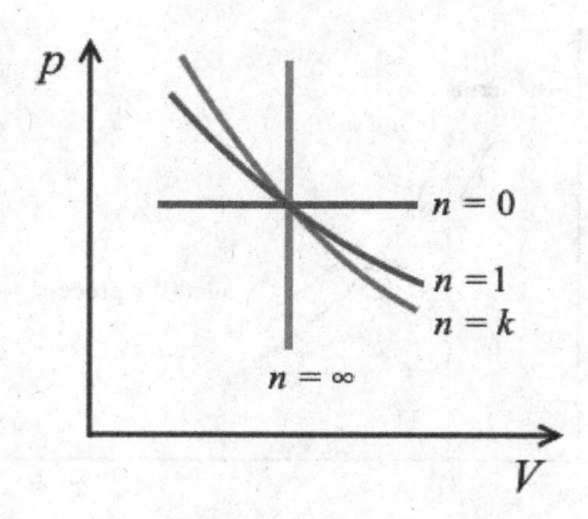

Fig. 4.7

- $n = 1$, results in $pv =$ cst, which is an isothermal process for a perfect gas;
- $n = k$, which is a reversible adiabatic process for a perfect gas.

The work associated with the polytropic process (for one unit of mass) is

$$\begin{aligned} w &= \int_1^2 p\,dv = \int_1^2 \frac{p_1 v_1^n}{v^n} dv \\ &= p_1 v_1^n \int_1^2 \frac{dv}{v^n} = \frac{p_1\, v_1^n (v_2^{1-n} - v_1^{1-n})}{1-n} \\ w &= \frac{p_2 v_2^n v_2^{1-n} - p_1 v_1^n v_1^{1-n}}{1-n} = \frac{p_2 v_2 - p_1 v_1}{1-n} \end{aligned} \tag{4.22}$$

The heat exchanged:

$$q - \frac{p_2 v_2 - p_1 v_1}{1-n} = u_2 - u_1 \tag{4.23}$$

Non-flow thermodynamic processes of an ideal gas — problems

4.1. The measured tire pressure at the beginning of a trip is 2.81 bar when the tires are in thermal equilibrium with the ambient at 284 K. At the end of the trip, the gauge shows 3.01 bar. Neglecting the expansion of the tires, determine the air temperature inside the tires at the end of the trip.

4.2. During a process at constant pressure (1.01 bar), a system absorbs 1350 J of heat while its internal energy increases by 1150 J. Determine the change in volume of the system.

4.3. An ideal gas produces work isothermally. In the process, 5200 J of heat is absorbed by the gas. How much work does the gas do?

4.4. A piston–cylinder apparatus contains air at 1.2 bar and 1.25 m^3/kg. Weights are added gradually to the piston such that the air is compressed very slowly until it reaches the pressure of 5 bar. Determine: (a) the temperature of the process; (b) the work associated with the process; (c) the heat exchanged with the surroundings; (d) the change in enthalpy during the process.

4.5. One kilogram of nitrogen expands from state 1 — where $p_1 = 6$ bar and $v_1 = 0.25$ m^3/kg — to state 2 — where the volume increased 2.5 times. The expansion process is polytropic, with $n = 1.5$. Determine the work associated with the process 1–2.

Chapter 5

Flow Thermodynamic Processes

5.1. The steady-flow energy equation

Consider an open system[a] exchanging mass, heat and work with the surroundings at a uniform rate. The boundary of the system is known as *control surface* and the system itself as *control volume* (Fig. 5.1). A steady-flow has the following characteristics:

- no property at any given location within the system boundary changes with time. It also means that, during the entire steady-flow process, the total volume V of the system, the total mass m of the system and the total energy content E of the system remain constant;
- no property at an inlet to the open system or at an exit changes with time. Therefore during a steady-flow process, the mass flow rate, the energy flow rate, pressure, temperature, specific volume, specific internal energy, specific enthalpy, and the velocity of flow at an inlet or at an exit remain constant;
- the rates at which heat and work are transferred across the boundary of the system remain unchanged;

[a]In a closed system, changes of state occur *in time*, i.e., system properties change slowly (for the system to remain in equilibrium) and uniformly throughout the system; in an open system, changes of state occur *in space*, i.e., system properties change from point to point within the control volume, but at any point they remain constant during the process.

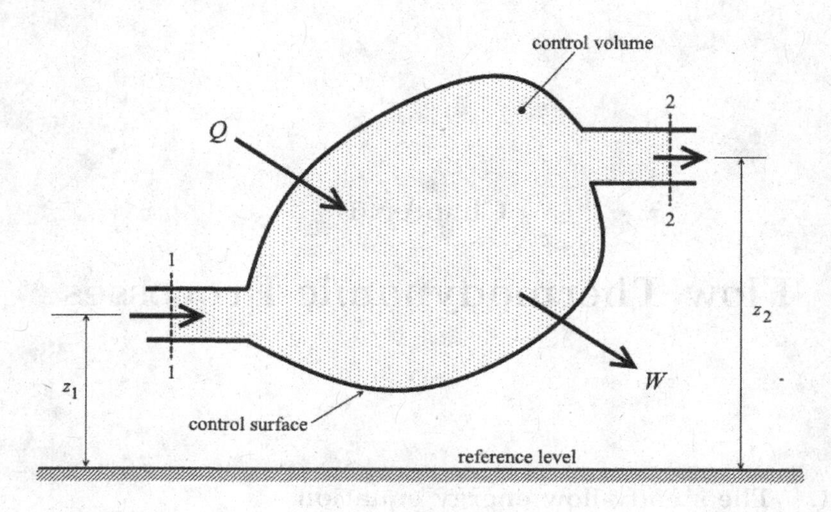

Fig. 5.1

- the mass flow rate through the system must be constant $\dot{m}_1 = \dot{m}_2$ (according to the equation of continuity).

Therefore, during any given time interval, the same amount of mass m enters and leaves the system. A steady-flow process can be regarded as a series of non-flow processes undergone by an imaginary closed system [1, pp. 37–38; 11, pp. 228–234; 7, pp. 96–98]. The energy associated to the mass m flowing through the system can be written as

$$W_{\text{flow}} = mFl = m\,pAl = m\,pv \tag{5.1}$$

where A is the cross-sectional area corresponding to a unit of mass (1 kg) of fluid being pushed into an already pressurized (p) volume for a length l. This is sometimes called **flow work** or **flow energy**; it should not be confused with the work W done on or by the open system.

The energy balance equation (3.7) for steady flow ($\Delta E_{\text{system}} = 0$) can be written as

$$\begin{aligned} E_{\text{in}} &= E_{\text{out}} \\ E_{\text{in}} &= Q + W_{\text{flow.in}} + E_{i.\text{in}} + E_{k.\text{in}} + E_{p.\text{in}}; \\ E_{\text{out}} &= W + W_{\text{flow.out}} + E_{i.\text{out}} + E_{k.\text{out}} + E_{p.\text{out}} \end{aligned}$$

Here E_i represents the internal energy of the fluid, E_k its kinetic energy and E_p its potential energy. Therefore,

$$Q + m\,p_1 A_1 l_1 + m u_1 + m\frac{V_1^2}{2} + mgz_1$$
$$= W + m\,p_2 A_2 l_2 + m u_2 + m\frac{V_2^2}{2} + mgz_2$$

where V represents the velocity of the fluid:

$$Q + m\,p_1 v_1 + m u_1 + m\frac{V_1^2}{2} + mgz_1$$
$$= W + m\,p_2 v_2 + m u_2 + m\frac{V_2^2}{2} + mgz_2$$

$$Q + m(p_1 v_1 + u_1) + m\frac{V_1^2}{2} + mgz_1$$
$$= W + m(p_2 v_2 + u_2) + m\frac{V_2^2}{2} + mgz_2 \tag{5.2}$$

$$Q + m\,h_1 + m\frac{V_1^2}{2} + mgz_1 = W + m\,h_2 + m\frac{V_2^2}{2} + mgz_2$$

$$Q - W = m(h_2 - h_1) + m\left(\frac{V_2^2}{2} - \frac{V_1^2}{2}\right) + mg(z_2 - z_1)$$

For some specific applications, the change in kinetic and potential energy is negligible by comparison to the other terms in Eq. (5.2). Considering $\Delta E_k, \Delta E_p \cong 0$ one obtains

$$Q - W = m(h_2 - h_1) \tag{5.3}$$

or

$$q - w = h_2 - h_1 \tag{5.4}$$

The last three equations are expressions of the **steady-flow energy equation**. They can be also written for a constant flow rate $\dot{m}$.

5.2. Open systems with steady flow

A large class of devices of interest to engineers operate most of the time under steady-state condition. We will introduce them briefly and the steady-flow energy equation will be applied to these devices treating them more or less like black boxes.

5.2.1. *Boilers and condensers*

These are components of thermal power plants. A *condenser* is used to condense the exhaust steam from a steam turbine so that the resulting pure water (steam condensate) may be reused in the *boiler* (or steam generator) as feed water to produce steam again. (See also [1, p. 41; 7, pp. 110–111; 11, p. 244].)

Figure 5.2 illustrates a boiler; Fig. 5.3 illustrates a condenser.

In these open systems the change in kinetic and potential energy is negligible. The fluid only passes through the system exchanging heat with the surroundings; no work is done ($W = 0$). The ideal thermal

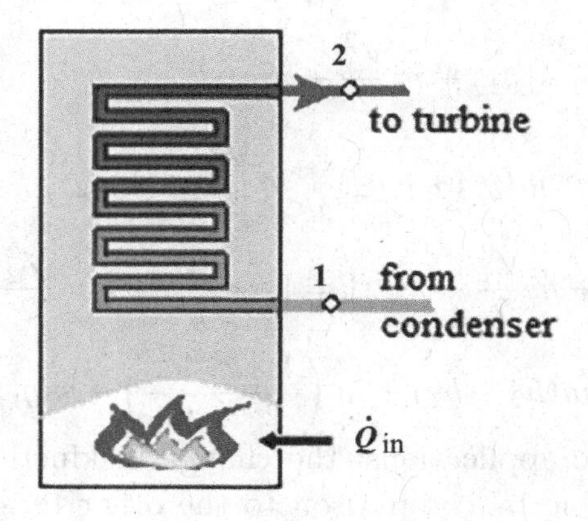

Fig. 5.2

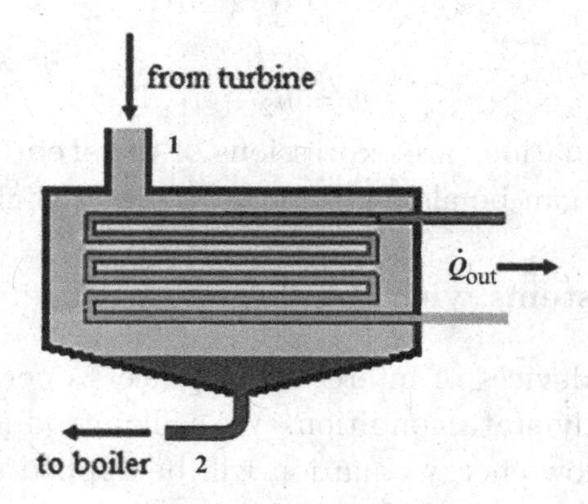

Fig. 5.3

processes occurring in these components are ideally isobaric. The steady-flow energy equation reduces to

$$Q = m(h_2 - h_1)$$

or

$$\dot{Q} = \dot{m}(h_2 - h_1) \tag{5.5}$$

Note that for the condenser, the cooling water is not part of the fluid of the open system but acts as a sink.

5.2.2. *Nozzles and diffusers*

A *nozzle* is a relatively short duct of variable cross-sectional area designed to accelerate a fluid by decreasing its pressure. The *diffuser* is designed to decelerate flow (see Fig. 5.4). Nozzles are used for various applications such as to increase the speed of the gases leaving the jet engine or rocket. Diffusers are used to slow down a fluid flowing at high speeds, such as at the entrance of a jet engine. The fluid flows very fast through these ducts, and the difference in kinetic energy cannot be neglected. Instead, one can consider the process adiabatic $Q = 0$. Also, no work is performed: $W = 0$. The change in potential energy is negligible [7, p. 104]:

$$m(h_2 - h_1) + m\left(\frac{V_2^2}{2} - \frac{V_1^2}{2}\right) = 0$$

or

$$\frac{1}{2}(V_2^2 - V_1^2) = (h_1 - h_2)$$

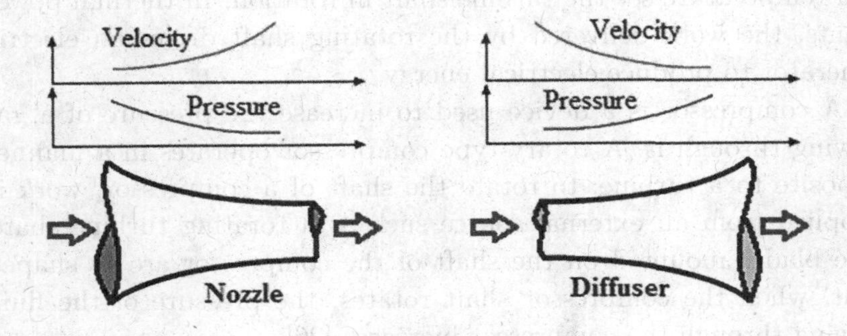

Fig. 5.4

After rearranging the terms,

$$h_1 + \frac{1}{2}V_1^2 = h_2 + \frac{1}{2}V_2^2$$
$$h_{01} = h_{02} \tag{5.6}$$

where

$$h_{01} = h_1 + \frac{1}{2}V_1^2$$

and

$$h_{02} = h_2 + \frac{1}{2}V_2^2$$

are called **stagnation** (or **total**) **enthalpies** [11, p. 850]. The stagnation enthalpy represents the enthalpy of a fluid when it is brought to rest adiabatically. In general, for any state of a fluid in steady-flow, one can define a (adiabatic reversible) **stagnation state**, where the kinetic energy is converted to enthalpy (internal energy plus flow energy). Based on Eq. (5.6) one can conclude that the stagnation enthalpy of a fluid remains constant during a steady-flow process.

5.2.3. *Turbines and compressors/pumps*

A turbine is a device with rows of blades mounted on a shaft which can be rotated about its axis (see Fig. 5.5 [25]). In steam turbines, steam at high pressure and temperature enters a turbine, sets the turbine shaft in rotation, and leaves at low pressure and temperature. In gas turbines, gaseous products of combustion at high pressure and temperature set the turbine shaft in rotation. In thermal power plants, the work delivered by the rotating shaft drives an electric generator to produce electrical energy.

A compressor is a device used to increase the pressure of a gas flowing through it. A rotary-type compressor operates in a manner opposite to a turbine: to rotate the shaft of a compressor, work is supplied from an external source such as a rotating turbine shaft. The blades mounted on the shaft of the compressor are so shaped that, when the compressor shaft rotates, the pressure of the fluid flowing through the compressor increases [26].

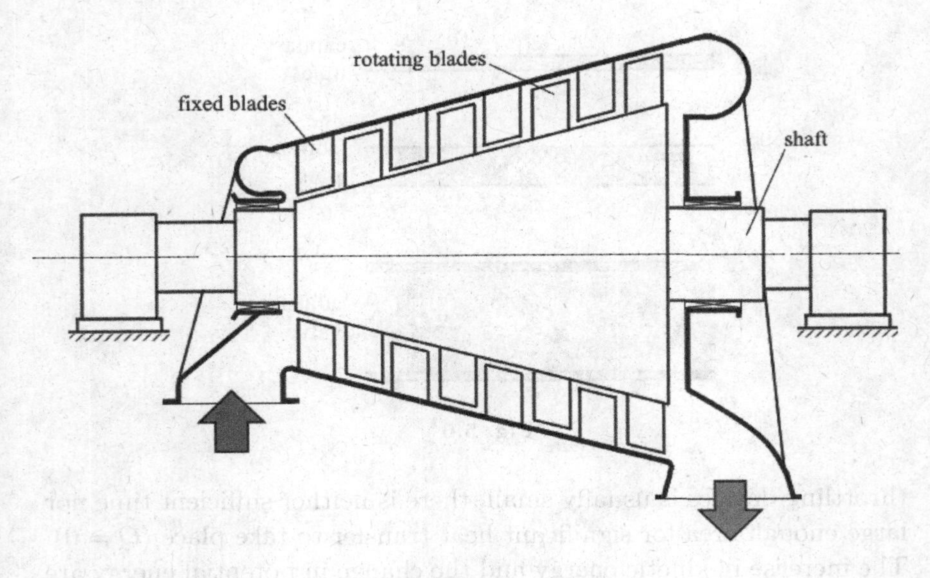

Fig. 5.5

A rotary pump works like a compressor except that it handles liquids instead of gases.

In these rotary devices the average velocity of the fluid is high but almost constant in magnitude. The change in kinetic and potential energy is negligible compared to the change in enthalpy level. Turbines, compressors and pumps are ideally operated under adiabatic conditions, so $Q = 0$. Under these conditions Eq. (5.2) reduces to:

$$W = m(h_1 - h_2) \tag{5.7a}$$

$$\dot{W} = P = \dot{m}(h_1 - h_2) \tag{5.7b}$$

5.2.4. *Throttling devices*

Throttling refers to a flow through some restriction with a relatively low speed. It does not involve any work ($W = 0$). A throttling device is used to produce a pressure drop in a flowing fluid. The device can be a partially open valve, a porous plug or a capillary tube placed in the path of the flowing fluid (see Fig. 5.6). Because the size of

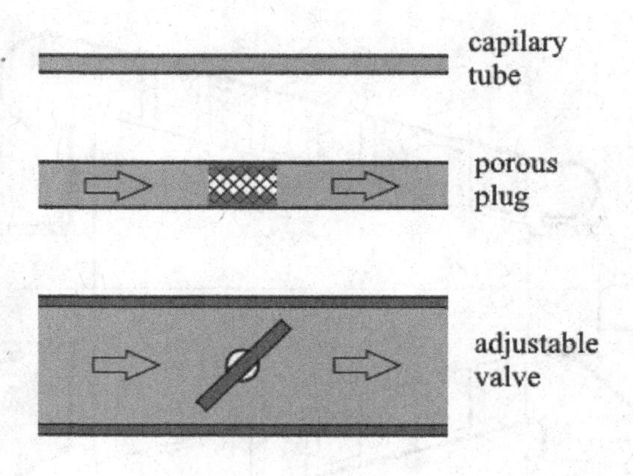

Fig. 5.6

throttling devices is usually small, there is neither sufficient time nor large enough area for significant heat transfer to take place ($Q = 0$). The increase in kinetic energy and the change in potential energy are insignificant. Therefore

$$h_1 = h_2 \tag{5.8}$$

5.3. Cycles consisting of steady-flow processes

Steady-flow processes are associated with the operation of distinct pieces of equipment as briefly described in Secs. 5.2.1–5.2.4. Each one of these components is treated as an open system. In order to obtain a thermodynamic cycle, a working fluid has to go from a component to the other in a closed circuit. When one extends the notion of system to the all the components in the circuit plus the fluid flowing through it, then during a steady-flow process the system itself does not, strictly speaking, undergo any process. This is because, if we define a process as a sequence of states of equilibrium occurring in time, all the properties of the system remain constant at every point within it. However, as it flows through the closed circuit, every particle of fluid is subjected to a series of distinct thermodynamic processes in such a way that, when the circuit is completed, the state of the fluid is identical to its state at the starting point. This is the case of the cyclic processes occurring in thermal power plants and refrigerators or heat pumps operating on a *closed cycle*, where the same fluid passes around the circuit continuously. There are, however, *open-cycle* plants, like gas turbine power plants, where the working fluid is taken from and rejected to the ambient, so the closing of the circuit involves the surroundings. In this case, special measures shall be taken for environmental protection.

In the case of internal combustion engines, all thermodynamic processes occur inside a cylinder considered a closed system, without any fluid flow.

More details about these cycles will be presented in Chapters 8–10.

Flow thermodynamic processes — problems

5.1. An ideal rotary compressor absorbs 50 m^3/min of normal atmospheric air at 15 °C and delivers it at 22 MPa. Determine the mass flow rate and the power required.

5.2. Combustion gases ($k = 1.38$) enter the nozzle of a gas turbine at 220 kPa, 160 °C, ($h_1 = 433$ kJ/kg) with a speed of 25 m/s and leave the nozzle at 100 kPa ($h_2 = 360.5$ kJ/kg) expanding adiabatically (quasi-static process). Determine the exit temperature and speed.

5.3. What is the physical interpretation of enthalpy?

Chapter 6

The Second Law of Thermodynamics

6.1. Cycle efficiency

According to the first law of thermodynamics, for a closed system, during any reversible change of state (process),

$$\delta Q - \delta W = dU$$

and, after integration, $\int \delta Q - \int \delta W = \int dU$, or

$$Q - W = \Delta U$$

For a closed path (cycle), an integral over it is a *circular integral*, or *cycle integral* (see also Sec. 3.3) and in this case, the circular integral of a state function is always zero. Therefore,

$$\oint dU = 0$$

Therefore $Q - W = 0$ or $Q = W$.

In this equation Q is the algebraic sum of all "heats" exchanged during the cycle considering the sign convention: $Q_{\text{in}} > 0$, $Q_{\text{out}} < 0$; also, W is the algebraic sum of all "works", $W_{\text{in}} < 0$, $W_{\text{out}} > 0$.

The first law becomes:

$$Q_{\text{in}} + Q_{\text{out}} = W_{\text{in}} + W_{\text{out}}$$

where Q_{in} is the heat supplied (positive), Q_{out} is the heat rejected (negative), W_{in} is the work done *on* the system (negative) while W_{out} is work done *by* the system (positive).

It follows that

$$Q_{\text{net}} = Q_{\text{in}} - |Q_{\text{out}}| = W_{\text{out}} - |W_{\text{in}}| = W_{\text{net}} \tag{6.1a}$$

Since in a cyclic process the amounts of heat and work are equal,

$$Q_{\text{net}} = W_{\text{net}}$$

we may have either

$$Q_{\text{net.in}} = W_{\text{net.out}} \quad \text{or} \quad Q_{\text{net.out}} = W_{\text{net.in}} \tag{6.1b}$$

(For simplicity, in the following sections the net work associated with the cycle will be denoted by W.)

If the net work is done *by* the system ($W_{\text{net.out}}$), the state changes happen in a *clockwise manner* and the cycle is called **power cycle** (see Fig. 6.1). A system operating in a power cycle is called a **heat engine**.

The performance index of a heat engine is called **thermal efficiency**, η_{th}. According to the general relation of definition (see Sec. 3.5.3), the thermal efficiency is the ratio of the net work output

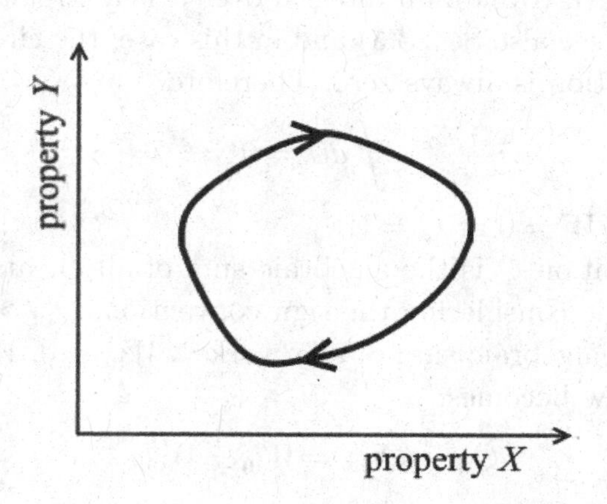

Fig. 6.1

to the total heat input:

$$\eta_{\text{th}} = \frac{W_{\text{net.out}}}{Q_{\text{in}}} \tag{6.2}$$

Both $W_{\text{net.out}}$ and Q_{in} are positive quantities, according to the sign convention. Therefore $\eta_{\text{th}} > 0$.

$$\eta_{\text{th}} = \frac{Q_{\text{in}} - |Q_{\text{out}}|}{Q_{\text{in}}} = 1 - \frac{|Q_{\text{out}}|}{Q_{\text{in}}} \tag{6.3}$$

Equation (6.3) shows that the greater the proportion of heat supply converted into work, the better the engine. Quantities Q_{in} and Q_{out} can be expressed per unit mass of the system, or they can be expressed per unit time

$$\eta_{th} = 1 - \frac{|\dot{Q}_{\text{out}}|}{\dot{Q}_{\text{in}}} = 1 - \frac{|q_{\text{out}}|}{q_{\text{in}}}.$$

6.2. The second law of thermodynamics (Planck statement)

According to the first law, net work cannot be produced during a cycle without some supply of heat (perpetual motion machine of the first kind is impossible). No mention is made, however, about how much of the heat introduced in the cycle ($Q > 0$) can be converted to work ($W > 0$). This aspect is dealt with by the second law, according to which some heat must always be rejected during the cycle ($|Q_{\text{out}}| > 0$), and therefore the cycle efficiency is always smaller than unity. There are several ways in which the second law of thermodynamics can be stated.

The **Planck**[a] **statement** of the second law is expressed as follows: *it is impossible for any device that operates on a cycle to receive heat from a single reservoir while producing a net amount of work* [1, p. 51; 7, p. 125; 11, p. 291]. This is also known as the *Kelvin–Planck statement.*

Consequently, for a system connected to a single reservoir the work produced by the cycle $W \leq 0$.

[a] Max Planck (1858–1947), German physicist.

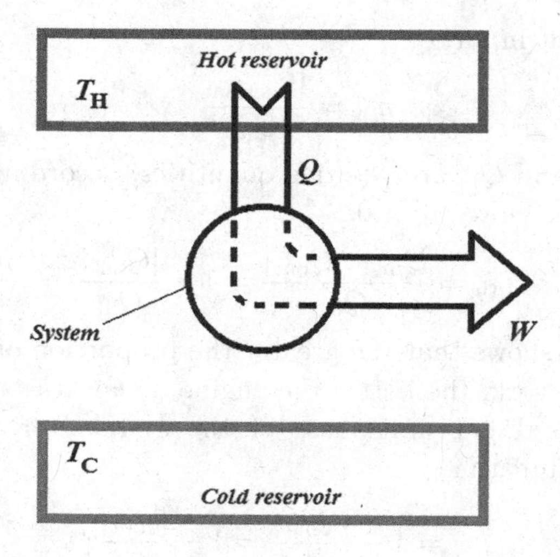

Fig. 6.2

This implies that if a system undergoes a cycle and produces work, it must operate between at least two reservoirs of different temperature, a reservoir of higher temperature than that of the fluid (from which it receives heat), and a reservoir of lower temperature (to which it rejects heat). A machine which will produce work continuously, while exchanging heat with only a single reservoir is known as a **perpetual motion machine of the second kind** (see Fig. 6.2).

The second law cannot be deduced from the first law; it is a separate axiom that represents a generalization from an enormous amount of observation. Its validity consists of the fact that neither the law, nor its consequences have ever been disproved by experience [1, p. 53].

Again, based on the second law, heat cannot be entirely extracted from a body and turned into work; therefore a heat engine can never have 100 % efficiency.

All the comments made so far implicitly referred to cycles describing the operation of heat engines. One can also devise a cycle during which heat is supplied at low temperature and rejected at high temperature, while work is done on the system (Fig. 6.3). This is

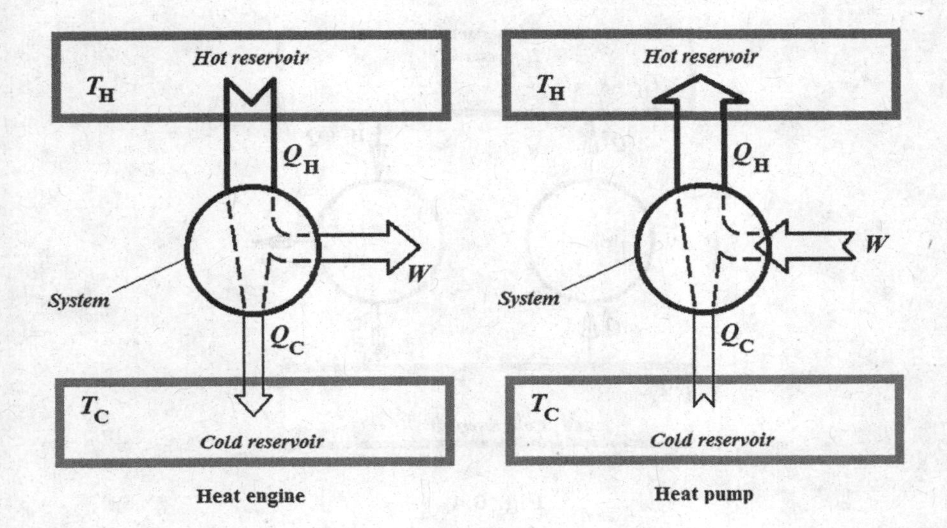

Fig. 6.3

called **refrigeration cycle** (or *heat pump cycle*) and during it the state changes happen in an *anticlockwise manner*.

6.3. Corollary of the second law (Clausius statement)

This first corollary, formulated by Clausius,[b] is often used as a statement of the second law: *it is impossible to construct a device that operates in a cycle and transfers heat from a cooler to a hotter body without work being done on the system by the surroundings* [1, p. 53; 7, p. 125; 11, p. 296]. Both, Planck and Clausius statements are equivalent.

Proof. The same technique known as "reductio ad absurdum" will be used; i.e., assume that the opposite of Clausius statement is true. This would be the case of a system (heat pump) for which the work $W = 0$ (see Fig. 6.4) while heat Q is transferred from a cold to a hot reservoir (based on the conservation of energy principle). A heat engine could

[b]Rudolf Julius Emanuel Clausius (born Rudolf Gottlieb, 1822–1888), German physicist and mathematician credited with making thermodynamics a science.

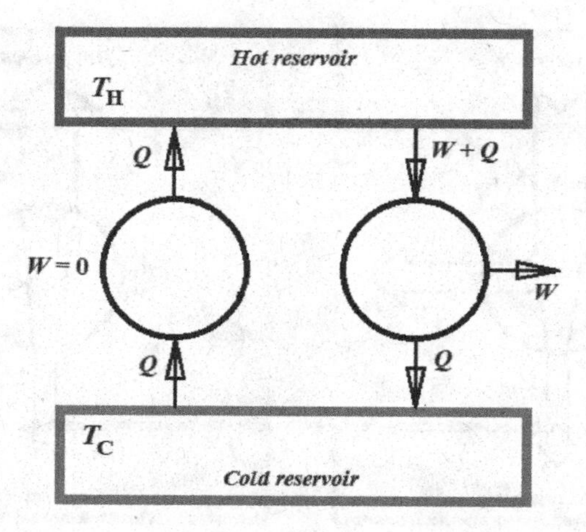

Fig. 6.4

operate in parallel between the same two reservoirs, while producing some net work W and rejecting heat Q to the cold reservoir. For this, according to the first law, the engine must receive heat in the amount $(W+Q)$ from the hot reservoir. The combined plant (heat pump plus heat engine) appears to be able to operate without the cold reservoir, since heat Q can be directly transferred to the heat pump, bypassing the reservoir. This combined plant would then be a system working in a cycle while extracting a positive amount of energy equal to $(W + Q) - Q = W$ from one single source and producing the net amount of work W. This is impossible according to the second law. Consequently, the original proposition must be true.

6.4. Reversible and irreversible processes

We have seen previously (Sec. 4.1) what a reversible process is. An alternative definition would be: *a reversible process is a process that can be reversed without leaving any trace on the surroundings.* Processes that are not reversible are called *irreversible processes.* The main point in this definition of reversible process is "restoration without leaving any trace on the surroundings" [1, p. 54; 7, p. 126; 11, p. 300]. That is, both the system and the surroundings are returned to their initial states at the end of the reverse process. This is possible only if the net heat and net work exchange between the system and the surroundings are zero for the combined process (original and reverse). A system can be restored to its initial state following a process, regardless of whether the process is reversible or irreversible. But for reversible process, their restoration is made without leaving any net change on the surroundings.

The idea of reversibility can be applied to non-flow as well as steady-flow processes; e.g., a fluid expanding adiabatically in a turbine and being compressed adiabatically in a rotary compressor. When reversible processes are used, work-producing devices (turbines, internal combustion engines) deliver the most work, while work-consuming devices (pumps, compressors) require the least work.

Unfortunately, reversible processes are only idealizations. All natural processes are irreversible to some extent. Despite of that, the study of reversible processes is important for two reasons:

(a) they are easy to analyze, and
(b) they serve as ideal models showing the theoretical limits for the corresponding real processes, like the ones occurring in an engine, a refrigerator, etc.

Also, there are some engineering situations where the effect of irreversibility can be neglected and the reversible process furnishes an excellent approximation to reality [27].

Phenomena that cause irreversibility are referred to as **irreversibilities**. The most common irreversibilities associated with thermodynamic processes are:

- *Friction.* When a gas inside a piston–cylinder apparatus expands and friction develops at the interface between piston and cylinder, part of the work produced is converted into heat and interface is getting hotter due to friction. When the motion of the piston is reversed (compression), the heat produced during expansion is not converted back into work as the piston is restored to its original position. Instead, more heat is generated due to friction. Therefore any process involving friction is irreversible. A similar phenomenon occurs during steady-flow processes in nozzles, turbines and rotary compressors (open systems).

- *Heat transfer through a finite temperature difference.* Consider a spontaneous heat transfer across a finite temperature difference from a hot reservoir to a cold one. To reverse the process and restore the initial state, heat should be pumped back to the hot reservoir without work input from any source. This is impossible since it violates the Clausius statement of the second law.
- *Unrestrained (free) expansion of a gas.* During this process (introduced in Sec. 4.1.1), $Q = W = 0$, $T =$ const. The only way to restore the system to its initial state is to compress it isothermally (reversibly) to its initial volume. This means that a certain amount of work is required from an external source and, from the first law, an equivalent amount of heat must also be rejected from the gas to the surroundings for the temperature of the gas to remain constant. At the end of the reversible compression the system has been returned to its initial state (no change in system state), the surroundings did work W on the system and received an amount of heat Q *equal* to W. However, the resaturation of the surroundings involves the conversion of that amount of heat *completely* into an equal amount of work, which would violate the second law. Therefore unrestrained expansion of gas is an irreversible process.
- *Mixing of gases.* The mixing process occurs spontaneously, but work should be supplied from the surroundings to separate the two gases. Hence mixing is an irreversible process.

When performing thermodynamic analyses it is convenient to separate irreversibilities into two categories [1, pp. 76–77; 7, p. 127; 11, pp. 302–303]:

- *internal irreversibilities* — that occur within the system, and
- *external irreversibilities* — that occur within the (immediate) surroundings.

In the case of *externally reversible processes*, no irreversibilities exist in the surroundings. Any heat transfer between the system and the surroundings occurs only across infinitesimal temperature difference. However, irreversibilities may exist within the system (e.g., internal friction, pressure gradients or eddies).

In the case of *internally reversible processes*, no irreversibilities exist within the system. During any process, the system goes slowly and without friction through a series of equilibrium states. However, irreversibilities may exist in the surroundings, usually due to heat transfer through a finite temperature difference.

6.5. Other corollaries of the second law

Corollary 2. *It is impossible to construct an engine operating between two heat reservoirs, which will have a higher efficiency than a reversible engine operating between the same two reservoirs* [1, p. 62].

Proof. Assume that the opposite of the statement is true. That is, there exists an engine that has a higher efficiency than a reversible engine operating between the same two reservoirs. Let X be that engine that operates between the same reservoirs as a reversible engine R, and assume that its efficiency $\eta_X > \eta_R$. Engine X receives heat Q_H from the source, produces work W_X, and rejects heat $(Q_H - W_X)$ to the sink (Fig. 6.5(a)). Reversible engine R receives the same heat Q_H from the same source as X and rejects to the same sink the heat $(Q_H - W_R)$ which is greater than $(Q_H - W_X)$. Because $\eta_X > \eta_R$, $W_X > W_R$.

If the reversible engine is reversed and operates as a heat pump (Fig. 6.5(b)), it receives heat $(Q_H - W_R)$ from the cold reservoir, receives work W_R from the surroundings and rejects heat Q_H to the hot reservoir. In this case, engine X can be coupled to the heat pump, covering the necessary work W_R, and producing a net work $(W_X - W_R)$. At the same time heat Q_H can be fed directly from the heat pump into engine X, bypassing the hot reservoir. It appears that the result is a combined power plant that produces a net amount of work $(W_X - W_R)$, while receiving heat $(Q_H - W_R) - (Q_H - W_X) = (W_X - W_R)$. This is impossible according to the second law, and the assumption that $\eta_X > \eta_R$

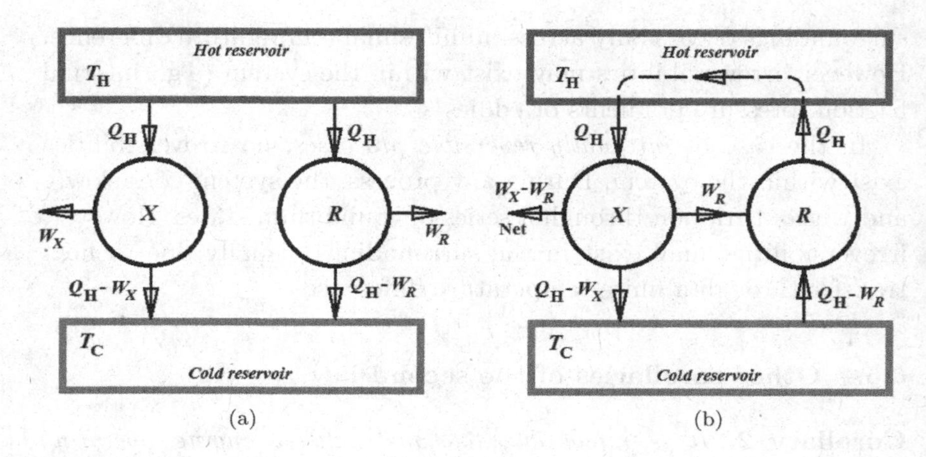

Fig. 6.5

is incorrect. Consequently, the original sentence must be true [1, pp. 62–63].

Corollary 3. *All reversible engines operating between the same two reservoirs have the same efficiency* [1, p. 63].

Proof. Once the second corollary has been demonstrated (by reductio ad absurdum), no special proof is required for this one. If no engine has a better efficiency than a reversible engine operating between the same reservoirs, it follows that all reversible engines have the same efficiency when operating under these conditions. This suggests that their efficiency must depend on the only parameter that may be different in every case, and that is the temperature of each reservoir. This is called the *Carnot efficiency*.

Corollary 4. *A scale of temperature can be defined which is independent of any particular thermometric substance, and which provides an absolute zero of temperature* [1, p. 63].

Proof. Let us consider a reversible engine that operates between two reservoirs as represented in Fig. 6.6. Corollary 3 implies that its efficiency depends only on the temperature of the reservoirs, and is,

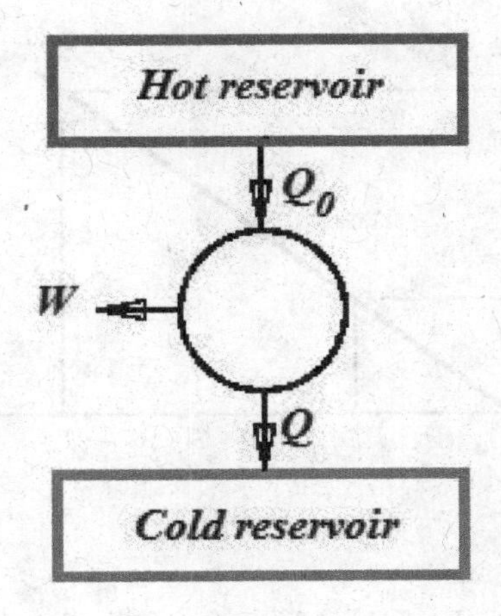

Fig. 6.6

implicitly independent of the nature of the working fluid and of the particular amount of heat supplied, Q_0. From Eq. (6.3)

$$\eta_{\text{th}} = \frac{Q_0 - |Q|}{Q_0} = 1 - \frac{|Q|}{Q_0} = 1 - \frac{T_C}{T_H} \tag{6.4}$$

and therefore $\frac{|Q|}{Q_0} = \frac{T_C}{T_H}$. Now let us attribute some positive number T_0 to T_H, the temperature of a source that can be easily reproduced (for instance, a pure substance condensing at a definite pressure) and let us denote simply by T the temperature of any cold reservoir (T_C). Now T can be expressed as

$$T = T_0 \frac{|Q|}{Q_0} \tag{6.5}$$

This way, T is uniquely defined by T_0 (chosen arbitrarily) and the ratio Q/Q_0 which is fixed for two given reservoirs. If Q is measured for various sinks at different temperatures and plotted against T, the result will be a scale of temperature linear in Q (Fig. 6.7). The slope of the line is Q_0/T_0. The *unit of temperature* is $1/T_0$ of the interval between $T = 0$ and T_0. This way a new scale of temperature was defined, which is called *thermodynamic scale*, independent of any thermometric property of any substance. This scale depends only on the laws of thermodynamics. It is also called *absolute scale*

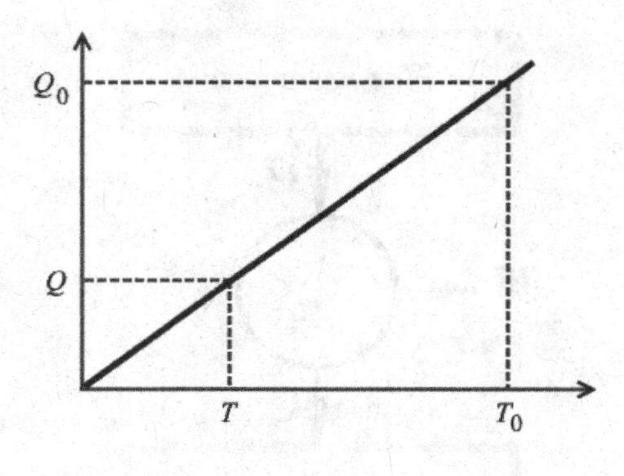

Fig. 6.7

because it is associated to the idea of an absolute zero: when $Q = 0$, $T = 0$. According to the second law of thermodynamics, Q — the heat rejected to the cold reservoir — can never be zero. This leads us to the idea that *absolute zero* is only a theoretical limit, which cannot be reached in practice. Indeed, this is stipulated by the **third law of thermodynamics**, also known as *Nernst*[c] *theorem*, concerned with the limiting behavior of systems as the temperature approaches zero. This theorem[d] implies the impossibility of attaining absolute zero through a finite number of thermodynamic processes, since as a system approaches absolute zero, the further extraction of energy from that system becomes more and more difficult [28].

From the ideal gas equation of state, $pV = mRT$ where p is the gas pressure, V is the volume of the container holding the gas, m is the mass of the gas filling the container, R is the gas constant, and T is the temperature measured on an absolute scale. For a fixed mass and volume of gas, the ideal gas law implies a direct proportionality between temperature and pressure:

$$p = \frac{mR}{V} T = \mathrm{cst} \cdot T \tag{6.6}$$

[c]Walther Hermann Nernst (1864–1941) was a German physicist.

[d]The third law has application mostly in chemical thermodynamics and low-temperature physics.

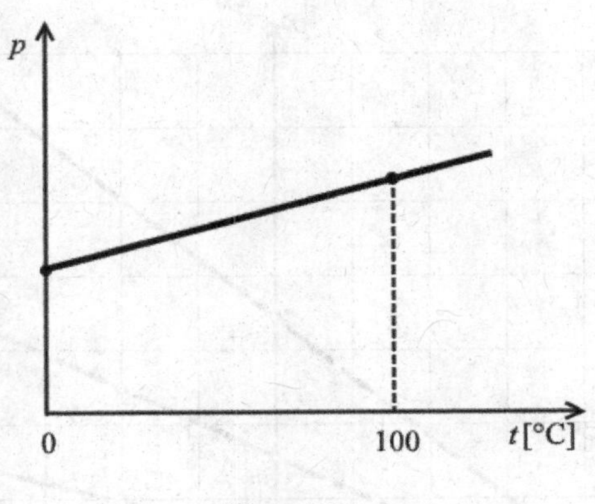

Fig. 6.8

where "cst" is a constant for a given m and V. Based on Eq. (6.6), a *constant-volume gas thermometer* [29, p. 334] can be calibrated such that it indicates the pressure in terms of temperature using two fixed points, e.g., the freezing point and the boiling point of water. These pressure and temperature values can be plotted on a graph as in Fig. 6.8.

The line connecting the two points serves as calibration curve for measuring unknown temperatures. Experiments show that, if temperatures are measured with gas thermometers containing different gases, thermometer readings are almost independent of the type of gas used, as long as the gas pressure is low and the temperature is well above the point of liquefaction. If temperature measurements are repeated with a gas at different starting pressures at 0 °C, straight-line calibration curves will be generated each time, as long as the pressure is low (solid lines marked Trial 1, Trial 2 and Trial 3 in Fig. 6.9 [29, p. 325]). If these lines are extended below 0 °C, regardless of the type of gas used or the value of the low starting pressure, the calibration lines converge to the point of coordinates ($p = 0$, $t = -273.15$ °C). This fact suggests that this particular temperature does not depend on the nature of the gas used in the thermometer. Because $p = 0$ represents the lowest possible pressure (perfect vacuum), it was concluded that $t = -273.15$ °C is the lowest temperature attainable, an *absolute zero*. By international

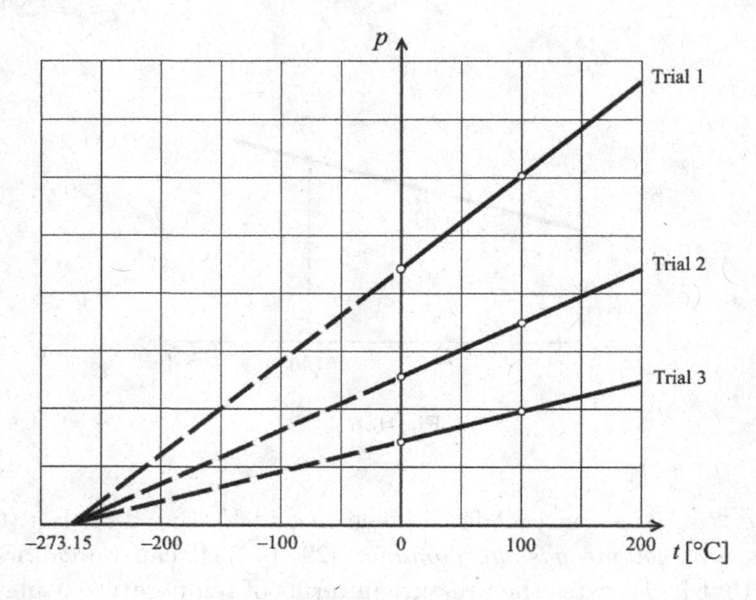

Fig. 6.9

agreement, the unit[e] on the thermodynamic (absolute) scale (1 K) was chosen to be identical to the unit on the Celsius scale (1 °C) and *absolute zero* (0 K) was set at exactly −273.15 °C.

Corollary 5. *Whenever a system undergoes a cycle,* $\oint \frac{\delta Q}{T}$ *is zero if the cycle is reversible and negative if irreversible, i.e., in general* $\oint \frac{\delta Q}{T} \leq 0$ [1, p. 67].

Proof. The approach described in [7] will be used. A system (the "System" in Fig. 6.10) receives energy δQ where its local temperature is T, as it produces work δW. This energy is received from a reservoir at T_{res} indirectly, by means of an intermediary system operating in a reversible cycle. This way no irreversibility is introduced as a result of heat transfer between the reservoir and the system. (This

[e] Actually, the 13th *Conférence générale des poids et mesures — CGPM* (*General Conference on Weights and Measures*) declared in Resolution 4 (1967) that "The kelvin, unit of thermodynamic temperature, is equal to the fraction 1/273.16 of the thermodynamic temperature of the triple point of water." The triple point of water is at 0.01 °C.

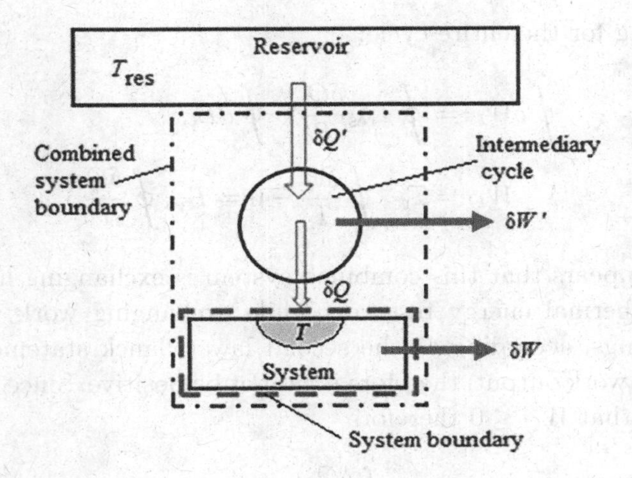

Fig. 6.10

arrangement allows also the possibility of heat transfer from the system to the reservoir.) The cycle receives energy $\delta Q'$ from the reservoir and supplies δQ (defined as magnitude) to the system while producing work $\delta W'$. Based on Eq. (6.4) one can write:

$$\frac{\delta Q'}{\delta Q} = \frac{T_{\text{res}}}{T} \Rightarrow \delta Q' = T_{\text{res}} \frac{\delta Q}{T} \tag{6.7}$$

As temperature T may vary, a multiplicity of such reversible cycles may be required.

Considering the combined system limited by the dotted line in Fig. 6.10 (the system plus the reversible system), one can write

$$dE_C = \delta Q' - \delta W_C \tag{6.8}$$

where dE_C denotes the change in energy of the combined system calculated as the difference between $\delta Q'$ and the total work of the combined system equal to $(\delta W + \delta W')$. Replacing in (6.8) $\delta Q'$ from (6.7) and solving for δW_C:

$$dE_C = T_{\text{res}} \frac{\delta Q}{T} - \delta W_C$$

$$\delta W_C = T_{\text{res}} \frac{\delta Q}{T} - dE_C$$

Integrating for the entire cycle:

$$\begin{aligned} \oint \delta W_C &= \oint T_{\text{res}} \frac{\delta Q}{T} - \oint dE_C \\ W_C &= T_{\text{res}} \oint \frac{\delta Q}{T} - 0 = T_{\text{res}} \oint \frac{\delta Q}{T} \end{aligned} \tag{6.9}$$

Since it appears that this combined system is exchanging heat with a single thermal energy reservoir while exchanging work with the surroundings, according to the second law (Planck statement) W_C cannot be work output; therefore it cannot be positive. Since $T_{\text{res}} > 0$, it follows that $W_C \leq 0$ therefore

$$\oint \frac{\delta Q}{T} \leq 0 \tag{6.10}$$

This is the **Clausius inequality**. It is valid for all thermodynamic cycles, reversible or irreversible, including refrigeration cycles [7, p. 141; 11, p. 339]. In the absence of any irreversibilities, the combined system would be an internally reversible one. If its cycle is reversed, all quantities will have the same magnitude but the opposite sign: work W_C cannot be positive for a power cycle; also, work cannot be negative for a reversed cycle. The only possibility is $W_{\text{C.int.rev}} = 0$. So

$$\oint \left(\frac{\delta Q}{T}\right)_{\text{int.rev}} = 0 \tag{6.11}$$

In conclusion, the equality in expression (6.10) holds for reversible cycles; the inequality holds for irreversible ones.

Corollary 6. *There exists a property of a closed system such that a change in its value is equal to* $\int_1^2 \frac{\delta Q}{T}$ *for any reversible process undergone by the system between states* 1 *and* 2 [1, p. 71].

Proof. Based on (6.11) we can conclude that $(\frac{\delta Q}{T})_{\text{int.rev}}$ is a property (or state function — because its cycle integral is zero). It is called

entropy and its symbol is $\boldsymbol{S}$:

$$\oint \left(\frac{\delta Q}{T}\right)_{\text{int.rev}} = 0 \tag{6.12}$$

For a process between states 1 and 2,

$$\int_1^2 \left(\frac{\delta Q}{T}\right)_{\text{int.rev}} = \Delta S = S_2 - S_1 \tag{6.13}$$

In differential form,

$$dS = \left(\frac{\delta Q}{T}\right)_{\text{int.rev}} \tag{6.14}$$

A more formal demonstration of the corollary would follow the same approach as for Corollary 1 of the first law (Sec. 3.5). Note that integrated form (6.13) defines the change in entropy rather than the entropy itself, just as the case of internal energy, where its change has been defined mathematically rather than the energy itself. Absolute values of entropy are defined based on the third law of thermodynamics, which says that for a perfect crystalline structure, at absolute zero (0 kelvin), the entropy (S) is zero. For practical reasons, explained in Sec. 2.3, the entropy of a substance can be assigned a zero value at some arbitrarily selected reference state, and the entropy at some other particular state is calculated based on Eq. (6.13) where 1 represents the reference state (where $S = 0$) and 2 is the state at which the entropy is to be determined.

Measuring units for entropy S are [J/K] or [kJ/K] in SI, and [Btu/°R] in US Customary system. For one unit mass of substance (1 kg) the *specific entropy* is used:

$$s = \frac{S}{m} \quad \text{and} \quad ds = \left(\frac{\delta q}{T}\right)_{\text{int.rev}}$$

It is not easy to give an intuitive image of entropy. However, one can say that entropy is a function that describes how many different energy levels are available in a thermodynamic system (this aspect is related to quantum mechanics). The more levels, the higher the entropy. It also describes the spontaneity of a process to occur, since natural processes are always associated to an increase of the entropy of the system along with that of its surroundings. Entropy is — in other words — a measure of probability for a certain process. Every system left to itself will, on the average, change toward a condition of maximum probability, therefore maximum entropy. Also, entropy is a measure of the unavailability of heat energy for work.

The concept of entropy went beyond the frontiers of thermodynamics: it is used in the context of information theory, one can talk about entropy of language, or entropy of proteins [30].

Corollary 7. *The entropy in a closed system which is thermally isolated from the surroundings either increases or, if the process undergone by the system is reversible, remains constant* [1, p. 72].

Proof. When a closed system is thermally isolated from the surroundings, it will behave adiabatically ($\delta Q = 0$). Two situations are to be considered here.

First we will look at the simple case of a reversible process. This means that $dS = (\delta Q/T)$. Consequently, for any reversible adiabatic process the change in entropy is zero. For this reason, a process that is adiabatic and reversible is called an **isentropic process**.

The second situation is when an irreversible process A occurs between two states, 1 and 2 (Fig. 6.11). In this case there must be some change in the entropy of the system, since equation (6.14) does

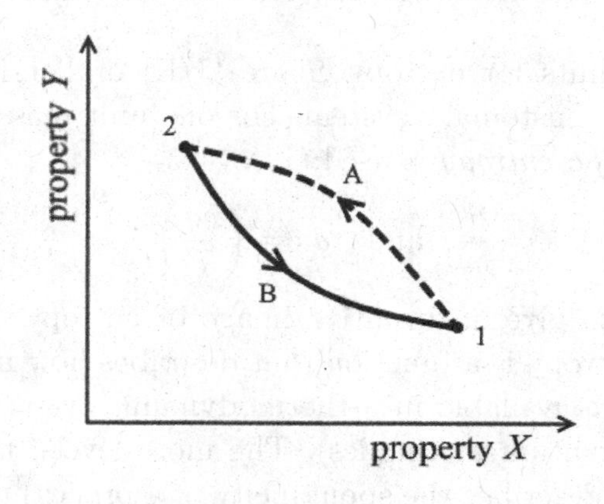

Fig. 6.11

not apply. However, the process being adiabatic, $\delta Q = 0$; therefore

$$\int_1^2 \left(\frac{\delta Q}{T}\right)_A = 0$$

Assume now that the system returns to the initial state by some reversible process B, not necessarily adiabatic. For this one can write

$$\int_2^1 \left(\frac{\delta Q}{T}\right)_B = S_1 - S_2$$

The whole cycle $1A2B1$ is irreversible since it contains one irreversible process. Hence, from Corollary 6

$$\oint \frac{\delta Q}{T} = \int_1^2 \left(\frac{\delta Q}{T}\right)_A + \int_2^1 \left(\frac{\delta Q}{T}\right)_B = \int_2^1 \left(\frac{\delta Q}{T}\right)_B < 0$$

Therefore $(S_1 - S_2) < 0$ or $S_2 > S_1$.

6.6. The Carnot cycle

The Carnot cycle consists of two adiabatic and two isothermal processes. All these four processes are totally reversible (internally and externally). The cycle (Fig. 6.12) was devised by Nicolas Léonard

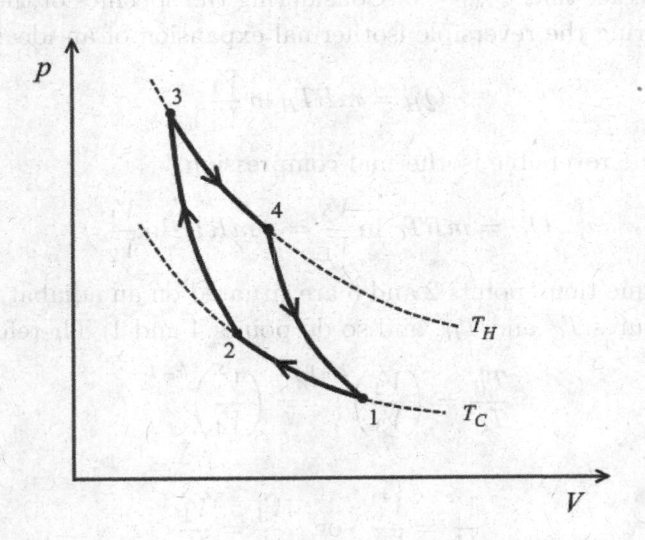

Fig. 6.12

Sadi Carnot in the 1820s. In a heat engine using the Carnot cycle the working fluid (an ideal gas) is compressed isothermally (1–2) at T_C while rejecting heat Q_C to a cold sink; then it is compressed adiabatically (2–3) until it reaches temperature T_H. Then the fluid expands first isothermally (3–4) at T_H, while receiving heat Q_H from a hot source, then adiabatically (4–1) until it reaches again the temperature T_C.

The thermal efficiency of any power cycle, including the Carnot cycle, can be expressed as

$$\eta_{\text{th}} = \frac{W_{\text{net.out}}}{Q_{\text{in}}}$$

With the notations in Fig. 6.12,

$$\eta_{\text{th}} = \frac{W_{34} + W_{41} + W_{12} + W_{23}}{Q_{\text{in}}}$$

Also, from Eq. (6.3)

$$\eta_{\text{th}} = \frac{Q_{\text{in}} + Q_{\text{out}}}{Q_{\text{in}}} = \frac{Q_{\text{in}} - |Q_{\text{out}}|}{Q_{\text{in}}} = 1 - \frac{|Q_{\text{out}}|}{Q_{\text{in}}} = 1 - \frac{|Q_C|}{Q_H}$$

Remember that in this equation the negative sign takes into consideration the fact that $Q_{\text{out}} < 0$. Considering the specifics of the Carnot cycle, during the reversible isothermal expansion of an ideal gas

$$Q_H = mRT_H \ln \frac{V_4}{V_3}$$

During the reversible isothermal compression,

$$Q_C = mRT_C \ln \frac{V_2}{V_1} = -mRT_C \ln \frac{V_1}{V_2}$$

At the same time, points 2 and 3 are situated on an adiabat between temperatures T_C and T_H, and so do points 4 and 1. Therefore:

$$\frac{T_H}{T_C} = \left(\frac{V_2}{V_3}\right)^{k-1} = \left(\frac{V_1}{V_4}\right)^{k-1}$$

Thus:

$$\frac{V_2}{V_3} = \frac{V_1}{V_4} \quad \text{or} \quad \frac{V_4}{V_3} = \frac{V_1}{V_2}$$

Consequently,

$$\frac{|Q_C|}{Q_H} = \frac{T_C}{T_H}$$

Therefore, for the Carnot cycle,

$$\eta_{\mathrm{th},C} = 1 - \frac{T_C}{T_H} \tag{6.15}$$

As explained at the beginning of this section, the Carnot cycle is totally reversible: it has no internal irreversibilities, when heat exchange with the surroundings takes place (processes 1–2 and 3–4) the temperature is constant and when temperature varies, there is no heat exchange ($Q = 0$ during processes 2–3 and 4–1).

Equation (6.15) — often referred to as the *Carnot efficiency* — shows that the thermal efficiency of this reversible cycle depends only on the temperature limits between which it evolves, and does not depend on the nature of the working fluid.[f] This corresponds to the conclusions of Corollary 3 as presented in Sec. 6.5. Also, temperatures T_C and T_H are thermodynamic (absolute) temperatures, which do not depend on a thermometric property of any substance.

The importance of the Carnot cycle consists of the fact that it has the highest efficiency of all heat engines operating between two reservoirs at given temperatures. In many situations we will refer to this cycle for comparison, since it is the best known reversible cycle.

The Carnot cycle can be executed in a closed system or in a steady-flow system, as will be discussed in later sections.

Being a totally reversible cycle, all its processes can be reversed, creating a **Carnot refrigeration cycle**. In this case the cycle remains the same, only occurring counterclockwise, the direction of each process being reversed (Fig. 6.13).

The enclosed area on the $p - V$ diagram represents the net work involved. Any cycle can be also represented in a $T - s$ diagram. The characteristic of such a diagram is that the area of the surface limited by the cycle corresponds to the net heat used in the cyclic process. Certainly, according to the first law (Eq. (6.1a)),

---

[f] The fluid can be an ideal gas or a condensable vapor.

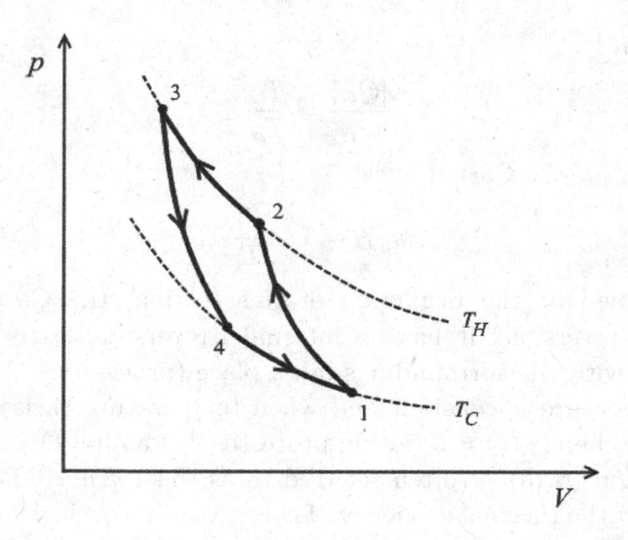

Fig. 6.13

$Q_{\text{net}} = W_{\text{net}}$. The direction of the cycle stays the same, regardless of the type of diagram: clockwise if the system does work ($W_{\text{net}} > 0$), counterclockwise if work is done on the system every cycle ($W_{\text{net}} < 0$). The $T - s$ diagram of a Carnot cycle is presented in Fig. 6.14. It is easy to observe from this diagram that

$$\eta_{\text{th},C} = 1 - \frac{|Q_{\text{out}}|}{Q_{\text{in}}} = 1 - \frac{|Q_C|}{Q_H} = 1 - \frac{mT_C(s_1 - s_2)}{mT_H(s_4 - s_3)} = 1 - \frac{T_C}{T_H}$$

6.7. Processes in T–s diagrams

As discussed in the previous section, with the introduction of entropy, we can now represent states and processes using a T–s diagram. One characteristic of this diagram is that the area under a process curve is related to the amount of heat exchanged by the gas. Also, T–s diagrams allow the visualization of the entropy changes during reversible or irreversible processes.

As exemplified for the case of the Carnot cycle, isothermal and isentropic (adiabatic reversible) processes are, by definition, horizontal and vertical lines, respectively. Less obvious are the lines of constant pressure and (specific) volume shown in Fig. 6.15. One

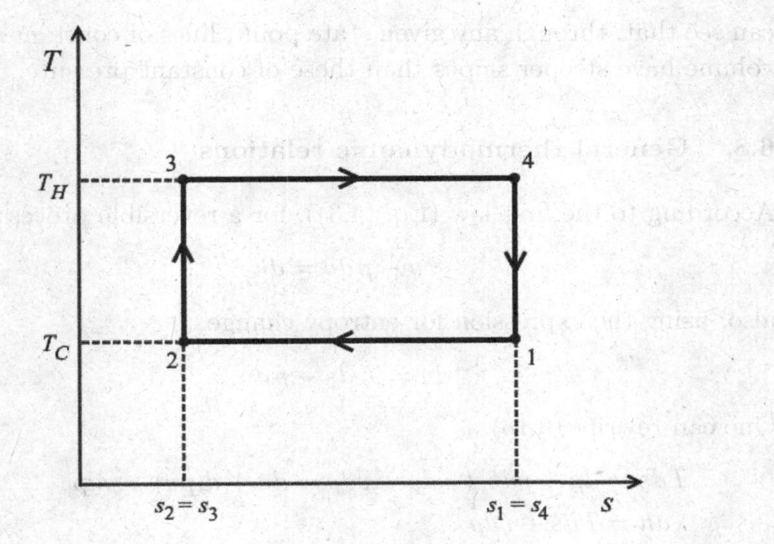

Fig. 6.14

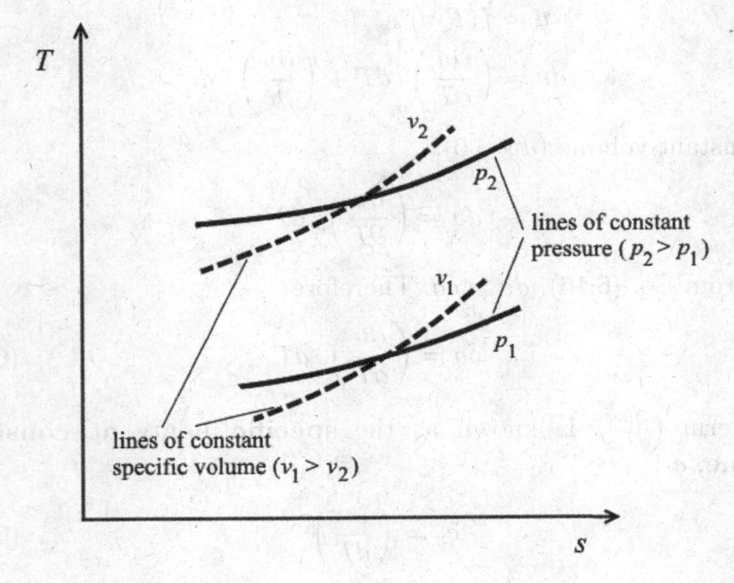

Fig. 6.15

can see that, through any given state point, lines of constant specific volume have steeper slopes than those of constant pressure.

6.8. General thermodynamic relations

According to the first law (Eq. (4.3)), for a reversible process:

$$\delta q - p\,dv = du$$

also, using the expression for entropy change,

$$du = T\,ds - p\,dv \tag{6.16}$$

One can rewrite (6.16) as

$$\begin{aligned} Tds &= du + pdv + vdp - vdp = du + d(pv) - vdp \\ dh &= Tds + vdp \end{aligned} \tag{6.17}$$

Any thermodynamic property can be expressed in terms of two independent intensive properties. From (6.16) one can write that

$$\begin{aligned} u &= f(T, v) \\ du &= \left(\frac{\partial u}{\partial T}\right)_v dT + \left(\frac{\partial u}{\partial v}\right)_T dv \end{aligned}$$

At constant volume ($dv = 0$),

$$du = \left(\frac{\partial u}{\partial T}\right)_v dT$$

and, from Eq. (6.16), $du = \delta q$. Therefore

$$\delta q = \left(\frac{\partial u}{\partial T}\right)_v dT \tag{6.18}$$

The term $(\frac{\partial u}{\partial T})_v$ is known as the **specific heat**[g] **at constant volume**, c_v:

$$c_v = \left(\frac{\partial u}{dT}\right)_v \tag{6.19}$$

[g]The specific heat, in general, is also referred to as "specific heat capacity" [1, p. 96].

Equation (6.19) shows that, physically, the specific heat at constant volume can be viewed as the energy required to raise the temperature of a unit mass of substance by one degree at constant volume. Measuring units for c_v are: $[c_v]_{\text{SI}} = \text{J/(kg K)}$; $[c_v]_{\text{US Customary}} = \text{Btu/(lb °R)}$.

For an ideal gas c_v is a function of temperature only. From Eq. (6.18), after integration at constant volume one obtains

$$q = \Delta u = \int c_v(T)dT \quad \text{or} \quad Q = \Delta U = m \int c_v(T)dT \qquad (6.20)$$

At low pressures, real gases approach ideal gas behavior, so $c_v = c_v(T)$. The effect of temperature is negligible for monoatomic gases. For a perfect gas, $c_v = \text{cst}$. Although the integration in Eq. (6.20) is straightforward, to avoid laborious calculations when solving problems involving gas power cycles, we will consider, for simplicity, that $c_v = \text{cst} = c_{v@300\text{ K}}$ (see Chapter 9).

For constant specific heats,

$$q = \Delta u = c_v \Delta T \quad \text{or} \quad Q = \Delta U = mc_v \Delta T \qquad (6.21)$$

From Eq. (6.17) one can write that

$$h = \phi(T, p)$$

Therefore

$$dh = \left(\frac{\partial h}{\partial T}\right)_p dT + \left(\frac{\partial h}{\partial p}\right)_T dp$$

At constant pressure ($dp = 0$),

$$dh = \left(\frac{\partial h}{\partial T}\right)_p dT$$

and, from Eq. (6.17), $dh = \delta q$. Therefore

$$dh = \delta q = \left(\frac{\partial h}{\partial T}\right)_p dT \qquad (6.22)$$

The term $(\frac{\partial h}{\partial T})_p$ is known as the **specific heat at constant pressure**, c_p:

$$c_p = \left(\frac{\partial h}{\partial T}\right)_p \qquad (6.23)$$

Physically, the specific heat at constant pressure can be viewed as the energy required to raise the temperature of a unit mass of substance by one degree at constant pressure. Measuring units for c_p are the same as for c_v.

For an ideal gas c_p is a function of temperature only. From Eq. (6.22), after integration at constant pressure one obtains

$$q = \Delta h = \int c_p(T)dT \quad \text{or} \quad Q = \Delta H = m \int c_p(T)dT$$

At low pressures, real gases approach ideal gas behavior, so $c_p = c_p(T)$. The effect of temperature is negligible for monoatomic gases. For a perfect gas, $c_p = \text{cst}$. When solving problems involving gas power cycles, we will consider, for simplicity, that $c_p = \text{cst} = c_{p@300\text{ K}}$ (see Chapter 9).

For constant specific heats,

$$q = \Delta h = c_p \Delta T \quad \text{or} \quad Q = \Delta H = mc_p \Delta T \tag{6.24}$$

Differentiating the equation of definition of enthalpy (3.6b) one obtains:

$$dh = du + d(pv)$$
$$dh = du + RdT$$
$$c_p dT = c_v dT + RdT$$

and hence

$$c_p = c_v + R$$
$$c_p - c_v = R \tag{6.25}$$

This is known as the *Mayer's relation*, named after the German physician and physicist Julius Robert von Mayer (1814–1899) who derived it.

Equation (6.25) shows that, although c_p and c_v may vary with temperature, their difference is constant [1, p. 107]. This is an important relationship for an ideal gas since it makes possible the calculation of c_v, for example, when c_p and R are known.

Using Eq. (6.17) and dividing it by T, one obtains

$$\frac{dh}{T} = ds + \frac{v}{T}dp$$

Replacing dh by $c_p\, dT$, and v/T by R/p, the equation becomes

$$ds = \frac{c_p}{T} dT - \frac{R}{p} dp$$

By integrating between states (p_0, T_0) and (p, T),

$$\int_{s_0}^{s} ds = \int_{T_0}^{T} \frac{c_p}{T} dT - \int_{p_0}^{p} \frac{R}{p} dp$$

Considering $s_0 = 0$ at the reference state (p_0, T_0),

$$s = \int_{T_0}^{T} \frac{c_p}{T} dT - R \ln \frac{p}{p_0}$$

For constant c_p, the equation becomes

$$s = c_p \ln \frac{T}{T_0} - R \ln \frac{p}{p_0}$$

The change of entropy between some states 1 and 2 is

$$s_2 - s_1 = c_p \ln \frac{T_2}{T_1} - R \ln \frac{p_2}{p_1} \qquad (6.26)$$

The ratio of specific heats is k, the *adiabatic exponent*

$$k = \frac{c_p}{c_v} \qquad (6.27)$$

which is also known[h] as **specific heat ratio** [7, p. 77; 11, p. 183]. This ratio varies very mildly with temperature [11, p. 183].

For some selected gases, the values of ratio k can be found in Appendix E. At room temperature, many diatomic gases, including air, have a specific heat ratio of about 1.4.

The second law of thermodynamics — problems

6.1. At 55 °C and 1 bar, an amount of air occupies a volume of 1.5 m^3. Then the air is subjected to a sequence of processes as follows: isochoric heating to a pressure of 5 bar, followed by an isobaric increase in temperature up to 1750 °C, then an isothermal expansion until the volume reaches 2.5 m^3, ending

[h] In [1], k is referred to as *index of isentropic expansion* or *compression*.

with an isochoric cooling to the initial pressure. Determine: (a) the state of air (p, V, T) at the end of each process and represent the sequence of processes in p–V and $T-s$ diagrams; (b) the work and heat exchange and the variation of internal energy and enthalpy during each process, as well as during the entire sequence. Note: consider the following properties of air: $c_p = 1000$ J/(kg K), $R = 287$ J/(kg K).

6.2. When a real gas undergoes an adiabatic process, its entropy: (a) increases; (b) remains constant; (c) decreases.

6.3. When an ideal gas undergoes an adiabatic process, its entropy: (a) increases; (b) remains constant; (c) decreases.

6.4. According to the second law of thermodynamics: (a) the entropy of an isolated system can never decrease; (b) the entropy of an open system remains constant; (c) heat cannot be transferred spontaneously from a cold reservoir to a hot one.

6.5. A Carnot cycle consists of: (a) two isobars and two isentropes; (b) two isotherms and two isochores; (c) two isotherms and two adiabats.

6.6. In $T-s$ coordinates, a reversible Carnot cycle looks like: (a) a trapeze; (b) a rectangle; (c) curvilinear rectangle.

6.7. 1.5 kg of carbon dioxide undergoes an isochoric process 1–2 during which its pressure increases 1.8 times. Carbon dioxide is considered a perfect gas having $c_p = 825$ J/(kg K) and $R = 189$ J/(kg K). Determine the change in entropy for the system.

6.8. A Carnot cycle has an efficiency $\eta = 20$ % and the work done during the isothermal expansion is $W_{\text{iso.out}} = 100$ J. Find the work consumed during the isothermal compression.

6.9. A piston–cylinder apparatus contains 1 kg of air which undergoes a process from state 1 ($T_1 = 300$ K, $v_1 = 0.8$ m^3/kg) to state 2 ($T_2 = 420$ K, $v_2 = 0.2$ m^3/kg). Treating air as an ideal gas: (a) determine if this process is adiabatic and (b) if yes, determine the work W_{12} for the adiabatic process. If no, determine the direction of the heat transfer Q_{12}.

Chapter 7

Properties of Gas Mixtures: Moist Air

Air is a very common fluid in engineering thermodynamics. Its composition can be assumed invariable for most processes. Therefore, for practical reasons, air is usually treated as a single substance. However, for air-conditioning processes, atmospheric air has to be considered as a gas–water–vapor mixture. As long as no condensation occurs, both these constituents[a] can be considered ideal gases. It is, therefore necessary to have some way of deducing the properties of the mixture from those of its constituents.

7.1. Empirical laws for mixtures of gases

Two empirical laws are used to model the behavior of ideal-gas mixtures: *Dalton's law*[b] *of additive pressures* (or *law of partial pressures*) and *Amagat's law*[c] *of additive volumes* (or *law of partial volumes*). We will apply them to a mixture of k constituents.

Dalton's law of additive pressures: *The pressure of a mixture of gases is equal to the sum of the partial pressures of the individual*

[a]Individual gases in a mixture are called *components* or *constituents*.

[b]Formulated in 1801 by John Dalton (1766–1844), English chemist, physicist and meteorologist.

[c]Formulated in 1880 by Émile Hilaire Amagat (1841–1915), French physicist.

constituents. The partial pressure of a constituent is its pressure when it occupies a volume equal to the volume of the mixture at the temperature of the mixture

$$p = \sum_{1}^{k} p_i(V, T)$$

Amagat's law of additive volumes: *The volume of a mixture of gases is equal to the sum of the partial volumes of the individual constituents. The partial volume of a constituent is its volume when it is placed at the pressure and the temperature of the mixture*

$$V = \sum_{1}^{k} V_i(p, T) \tag{7.1}$$

In the equations above, k refers the total number of constituents.

Both these laws hold exactly for ideal-gas mixtures, but are not strictly followed by real gases for which deviations become quite large at high pressures.

When Dalton's law is extended to include also reference to internal energies, the proposition is referred to as the *Gibbs–Dalton law*: the pressure and internal energy of a mixture of gases are respectively equal to the sums of the pressures and internal energies of the individual constituents when each occupies a volume equal to that of the mixture at the temperature of the mixture [1, p. 289]:

$$p = \sum_{1}^{k} p_i(V, T) \quad \text{and} \quad U = \sum_{1}^{k} U_i(V, T) \tag{7.2}$$

In the above equations, the subscript i refers to a typical constituent i, and V and T refer to the properties of the mixture.

Based on the equation of definition of enthalpy, $H = U + pV$, the enthalpy of the mixture becomes

$$H = \sum_{1}^{k} U_i(V, T) + V \sum_{1}^{k} p_i(V, T)$$

But

$$H_i(V, T) = [U_i + p_i V](V, T) = U_i(V, T) + V p_i(V, T)$$

and therefore

$$H = \sum_{1}^{k} H_i(V, T) \tag{7.3}$$

If a mixture occupies a volume V at temperature T, the pressure p_i of any constituent when it alone occupies the entire volume V at temperature T(known as *partial pressure* of constituent i) is given by

$$p_i V = n_i \bar{R} T$$

where n_i is the amount of substance (number of moles) of the constituent i in the volume V. By summing up all the equations of state for each constituent, one obtains

$$V \sum_{1}^{k} p_i = \bar{R} T \sum_{1}^{k} n_i$$

The total number of moles for the mixture, n, is

$$n = \sum_{1}^{k} n_i$$

From Dalton's law we have

$$p = \sum_{1}^{k} p_i(V, T)$$

For the mixture, one can write

$$pV = n\bar{R}T$$

The molar mass of the mixture being M and its mass m,

$$M = \frac{m}{n}$$

Therefore

$$pV = \frac{m}{M} \bar{R} T = mRT$$

where $R = \bar{R}/M$ is the specific gas constant for the mixture. For a unit mass (1 kg) of mixture, we have

$$pv = RT$$

This shows that the equation of state for the mixture has the same form as that for any constituent.

One can write that

$$R = \frac{\bar{R}}{M} = \frac{n}{m}\bar{R}$$

Therefore, since $n = \sum_1^k n_i = \sum \frac{m_i}{M_i}$ and $M_i = \frac{\bar{R}}{R_i}$,

$$R = \frac{\bar{R}}{M}\sum_1^k m_i \frac{R_i}{\bar{R}} = \sum_1^k \frac{m_i}{m} R_i \tag{7.4}$$

The ratio m_i/m is the *mass fraction* of constituent i having the gas constant R_i. Because the equation of state of an ideal gas holds for the mixture, one can write also that

$$dU = mc_v dT$$

for the mixture as well as for each constituent. From Gibbs–Dalton law (7.2):

$$U = \sum_1^k U_i(V, T)$$

Since U is only a function of T,

$$mc_v dT = \sum_1^k m_i c_{vi} dT$$

Then, the specific heat for the mixture, c_v, is given by

$$c_v = \sum_1^k \frac{m_i}{m} c_{vi} \tag{7.5}$$

Once c_v and R for the mixture have been determined, for a perfect gas $c_p - c_v = R$ and therefore

$$c_p = \sum_1^k \frac{m_i}{m} R_i + \sum_1^k \frac{m_i}{m} c_{vi} = \sum_1^k \frac{m_i}{m} c_{pi} \tag{7.6}$$

It becomes evident that, to determine the change in u, h, and s for the mixture, the same procedure will be followed as for any ideal gas, by using the values for R, c_v and c_p obtained from Eqs. (7.4)–(7.6).

7.2. Gas and vapor mixtures: moist air (atmospheric air)

The atmospheric air is a mixture of gases (approximately 21 % O_2, 78 % N_2 and 1 % Ar and other gases, in volume) and water vapor (moisture). The study of moist air is important for heating, ventilation, and air-conditioning (HVAC). For these engineering applications, *atmospheric air* (moist air) is treated as a mixture of only two constituents: *dry air* and *water vapor*. In further notation we will use subscript a for properties referring to dry air, and subscript v for water vapor. All HVAC processes occur practically at constant (atmospheric) pressure, while temperature ranges approximately from −10 to +50 °C. Under these conditions air and water vapor can be treated as ideal gases with negligible error (under 0.2 % [11, p. 738]) and moist air can be treated as an ideal-gas mixture. According to Dalton's law, the ideal-gas mixture pressure p is

$$p = p_a + p_v \tag{7.7}$$

For the above-mentioned temperature range, the specific heat c_p of dry air can be considered constant, with an average value of 1.005 kJ/(kg K) (see [7, Table T-10]). Taking 0 °C as a reference temperature, which is a convenient approach for HVAC calculations, the enthalpy of dry air is

$$h_a = c_{pa}t = 1.005(\text{kJ/kgK})t(^\circ\text{C}) \quad [\text{kJ/kg}] \tag{7.8}$$

It is common practice in psychrometry and heat transfer to use degrees Celsius and kelvin simultaneously. This is possible because "the degree Celsius is the special name for the kelvin used to express Celsius temperatures. The degree Celsius and the kelvin are equal in size, so that the numerical value of a temperature difference or temperature interval is the same when expressed in either degrees Celsius or in kelvin" [31].

Most of the time, water vapor is found in atmospheric air in superheated state (i.e., above the saturation level), and this is the case for the majority of industrial processes involving moist air. When the partial pressure of the water vapor corresponds to the saturation pressure of water at the mixture temperature, p_g in Fig. 7.1, the

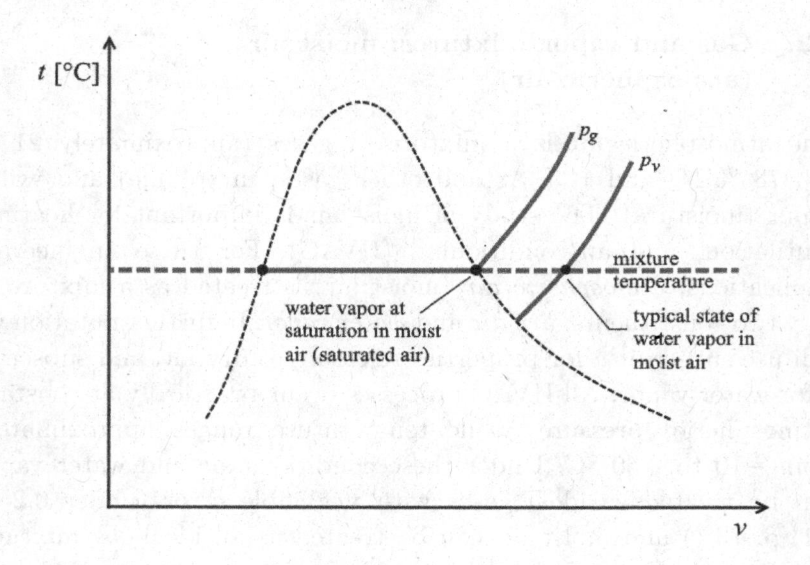

Fig. 7.1

mixture is said to be *saturated*. Because vapor (an ideal gas) can be considered as occupying alone the entire volume of the mixture, property tables for saturated and superheated steam can be used. At 50 °C, saturation occurs at 0.1235 bar (Appendix E, Table A.2). At this pressure and below it, water vapor can be treated as an ideal gas, with enthalpy depending only on temperature $h_v = h_v(T)$. This can be observed also from the T–s diagram of water (Appendix E, Fig. A.1): below 50 °C constant–enthalpy lines coincide with constant-temperature lines. Therefore, at constant temperature, the value of the enthalpy of superheated vapor can be approximated by the value for saturated vapor

$$h_v(t, \text{low}\, p) \approx h_g(t) \tag{7.9a}$$

This approach is used in the remainder of the section. Considering the enthalpy of saturated water $h_{f@0\,°\text{C}} = 0$, at some temperature t (°C) the enthalpy of water vapor is h_v:

$$h_g(t) \approx h_v(t) = h_{fg@0\,°\text{C}} + c_{pv}t(°\text{C}) \tag{7.9b}$$

Taking the latent heat of vaporization at 0 °C as $h_{fg} \approx 2501$ kJ/kg, and the average specific heat in the temperature range −10 to +50 °C as $c_{pv} \approx 1.88$ kJ/(kg K) (see [7, Table T-2 and Table HT-5]), we obtain the empirical expression for the enthalpy of water vapor

$$h_g(t) \approx h_v(t) = 2501 + 1.88\,t(^\circ\text{C}) \quad [\text{kJ/kg}] \tag{7.10}$$

Temperature-dependent expressions for the calculation of c_{pa} and c_{pv} can be found in [32].

7.2.1. *Properties of moist air*

Besides properties like internal energy, enthalpy, etc. that can be defined for any gas mixtures, some special properties can be defined for moist air. These properties are called *psychrometric properties* and the subject dealing with the behavior of moist air is known as *psychrometry*[d] or *hygrometry.* Also, because one unit mass of moist air can contain various amounts of water vapor, to eliminate any confusion, specific properties are referred to one unit mass of dry air, rather than one unit mass of moist air.

Absolute or **specific humidity** ω (also called *humidity ratio* or *moisture content*) is the ratio of the mass of water vapor in the mixture to the mass of dry air in the mixture, measured in kilograms of water vapor per kilogram of dry air (kg_v/kg_a).

$$\omega = \frac{m_v}{m_a} \quad (\text{kg}_v/\text{kg}_a) \tag{7.11}$$

Further,

$$\omega = \frac{m_v}{m_a} = \frac{p_v V/R_v T}{p_a V/R_a T} = \frac{p_v/R_v}{p_a/R_a} = \frac{R_a}{R_v}\frac{p_v}{p_a} = \frac{287}{461.5}\frac{p_v}{p_a} = 0.622\,\frac{p_v}{p_a}$$

therefore

$$\omega = 0.622\frac{p_v}{p - p_v} \quad (\text{kg}_v/\text{kg}_a) \tag{7.12}$$

where p is the barometric pressure.

When air is perfectly dry $p_v = 0$ and consequently $\omega = 0$. As more moisture is added, p_v and ω will increase, until the air can hold no more moisture. This limit corresponds to a condition called *saturated*

[d]The term derives from the Greek words — *psuchron* ($\psi\upsilon\chi\rho\acute{o}\nu$) — "cold" and *metron* ($\mu\acute{\varepsilon}\tau\rho o\nu$) — "measure".

air. For example, at t = 25 °C, (from [7, Table T-2, p. 522]) the saturation pressure of water vapor is $p_g = 0.03169$ bar = 3.169 kPa. This means that water vapor cannot exist in moist air of 25 °C at a higher pressure. Under normal atmospheric pressure (101.325 kPa) and a temperature of 25 °C, at saturation, the absolute humidity is

$$\omega_{\text{sat}} = 0.622 \frac{p_v}{p - p_v} = 0.622 \frac{3.169}{101.325 - 3.169} = 0.02008 \ \text{kg}_v/\text{kg}_a$$

If more moisture is introduced into saturated air, above ω_{sat}, excess water vapor will condense.

Relative humidity ϕ is the ratio of the actual mass of the water vapor m_v in a given volume of moist air to the mass of vapor m_g when the air is saturated at the same temperature.

$$\phi = \frac{m_v}{m_g} = \frac{p_v V / R_v T}{p_g V / R_v T} = \frac{p_v}{p_g} \tag{7.13}$$

Relative humidity is often expressed as a percentage. Saturation corresponds to $\phi = 100$ %.

From (7.11) and (7.12) it follows that

$$\phi = \frac{\omega p}{(0.622 + \omega) p_g} \tag{7.14}$$

and

$$\omega = \frac{0.622 \, \phi \, p_g}{p - \phi \, p_g} \tag{7.15}$$

The enthalpy of a parcel m of moist air is

$$H = H_a + H_v = m_a \, h_a + m_v \, h_v \tag{7.16}$$

Dividing (7.16) by m_a it follows that

$$h = \frac{H}{m_a} = h_a + \frac{m_v}{m_a} h_v = h_a + \omega \, h_v = h_a + \omega \, h_g \quad (\text{J/kg}_a) \tag{7.17}$$

since $h_v \approx h_g$.

Dew-point (temperature) t_{dp} is the temperature at which the condensation begins when the moist air is cooled at constant pressure.

If unsaturated moist air (state 1 in Fig. 7.2) is cooled at constant pressure p_v, the mixture will eventually reach the saturation

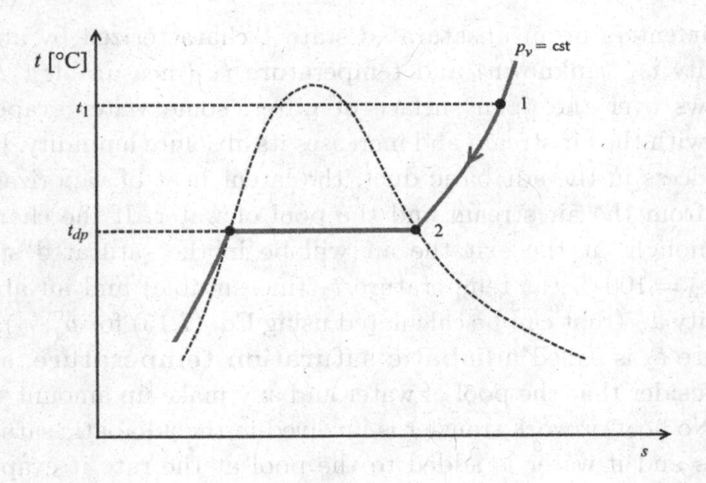

Fig. 7.2

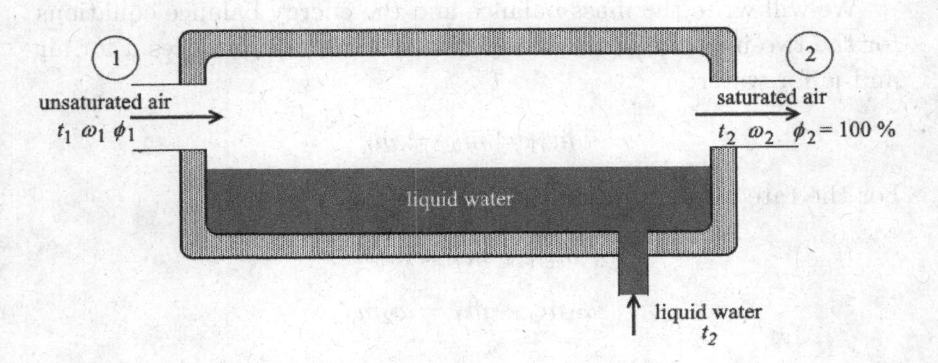

Fig. 7.3

temperature of water vapor (stated 2). This is t_{dp}; if the temperature drops any further, some vapor condenses out. This results in a reduction of absolute humidity ω and, consequently, a decrease of p_v (and p_g), while the relative humidity remains constant at 100 %.

Another way of determining the absolute and relative humidity is by means of an *adiabatic saturation process* shown schematically in Fig. 7.3. Consider a long insulated channel that contains a pool of water. A steady stream of moist air passes through the channel.

The air enters in an unsaturated state 1 characterized by absolute humidity ω_1 (unknown) and temperature t_1 (measurable). As the air flows over the open surface of water, some water evaporates, mixes with the air stream and increases its absolute humidity. During the process in the adiabatic duct, the latent heat of vaporization is taken from the air stream and the pool of water. If the channel is long enough, at the exit the air will be in the saturated state 2, with $\phi_2 = 100$ %, at temperature t_2 (measurable) and an absolute humidity ω_2 (that can be calculated using Eq. (7.15) for $p_{g@t2}$). Temperature t_2 is called **adiabatic saturation temperature**, and we will consider that the pool of water and any make-up amount will be at t_2. No heat or work transfer is involved in the adiabatic saturation process and if water is added to the pool at the rate it evaporates to maintain a constant liquid water level, the process is steady-state, steady-flow.

We will write the mass balance and the energy balance equations for the two-inlet, one-exit steady-flow system, using indices a for air and w for water:

$$\dot{m}_{a1} = \dot{m}_{a2} = \dot{m}_a$$

For the rate of evaporation $\dot{m}_f$, we have

$$\dot{m}_{w1} + \dot{m}_f = \dot{m}_{w2}$$

$$\omega_1 \dot{m}_a + \dot{m}_f = \omega_2 \dot{m}_a$$

or

$$\dot{m}_f = \dot{m}_a(\omega_2 - \omega_1)$$

For $\dot{Q} = 0$ and $\dot{W} = 0$, the energy balance yields $\dot{E}_{\text{in}} = \dot{E}_{\text{out}}$, or

$$\dot{m}_a h_1 + \dot{m}_a(\omega_2 - \omega_1) h_{f2} = \dot{m}_a h_2$$

Dividing by dry-air mass flow rate gives

$$h_1 + (\omega_2 - \omega_1) h_{f2} = h_2$$

or

$$(c_{pa} t_1 + \omega_1 h_{g1}) + (\omega_2 - \omega_1) h_{f2} = (c_{pa} t_2 + \omega_2 h_{g2})$$

which, solved for ω_1, yields

$$\omega_1 = \frac{c_{pa}(t_2 - t_1) + \omega_2(h_{g2} - h_{f2})}{h_{g1} - h_{f2}}$$

$$\omega_1 = \frac{c_{pa}(t_2 - t_1) + \omega_2 h_{fg2}}{h_{g1} - h_{f2}} \tag{7.18}$$

where

$$\omega_2 = \frac{0.622\, p_{g2}}{p - p_{g2}} \tag{7.19}$$

Equation (7.18) shows that the absolute humidity of air can be determined just by measuring the two temperatures t_1 and t_2; the other values can be easily found in thermodynamic tables. Once the value of ω is obtained, the relative humidity ϕ can be calculated. If air is already saturated when it enters the channel, temperatures t_1 and t_2 will be identical.

Dry-bulb temperature t_{db} is usually referred to as *air temperature* and represents the actual temperature of the air measured with a simple thermometer.

Wet-bulb temperature t_{wb} is the temperature at which liquid water, by evaporating into air, can bring the air to saturation at the same temperature. It is practically equal to the adiabatic saturation temperature; t_{wb} can be determined using a thermometer having a wetted sleeve of gauze snugly fit around its bulb (see Fig. 7.4). If a stream of unsaturated air flows past the thermometer, the evaporation from the wetted sleeve will result in a temperature fall and, when the thermometer reading reaches a steady value, that will be t_{wb}.

In practice, an instrument called *psychrometer* is used to measure the relative humidity in the atmosphere through the use of a dry-bulb and a wet-bulb thermometer. A basic type of device is the *sling psychrometer* [7; 11, p. 745], which is swung through the air in order to intensify the evaporation process. There are other, more complex devices, including *aspirating psychrometers*, which are equipped inside with spring-loaded or battery operated fans that ventilate the wet-bulb thermometer [7]. This creates an even evaporation rate, which produces a more accurate reading. In any case, the operation of these instruments is based on the experimental conclusion that although the wet-bulb temperature falls as the air

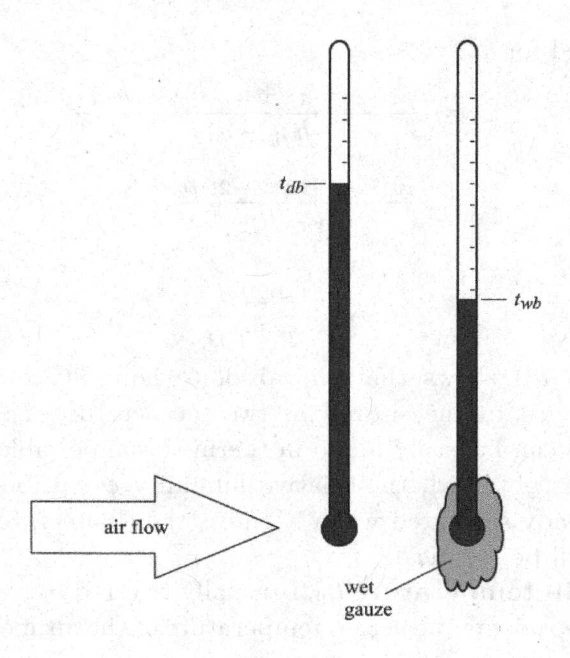

Fig. 7.4

velocity increases, it remains sensibly constant for air velocities between 2 and 40 m/s [1, p. 311]. Because Eq. (7.18) is quite complex, in practice the relation is expressed in the form of tables, where the relative humidity is indicated for a pair of values t_{db}, $(t_{\text{db}}-t_{\text{wb}})$.

7.2.2. *The psychrometric chart*

Calculations involving changes of state of moist air can be performed easily using the *psychrometric chart*. It represents (see Fig. 7.5) the properties of moist air in coordinates: dry-bulb temperature (t_{db}, as abscissa) and absolute humidity (ω, as ordinate). An enlarged version of this chart in SI units can be found in Appendix E.

This is the Anglo-American version (Carrier diagram). The European version of the psychrometric chart is the *Mollier diagram* (or Mollier chart [33]). They are identical in content but differ in appearance.

The state of a given parcel of moist air is represented by a point on the chart, known as the *status point*. If any two of the three

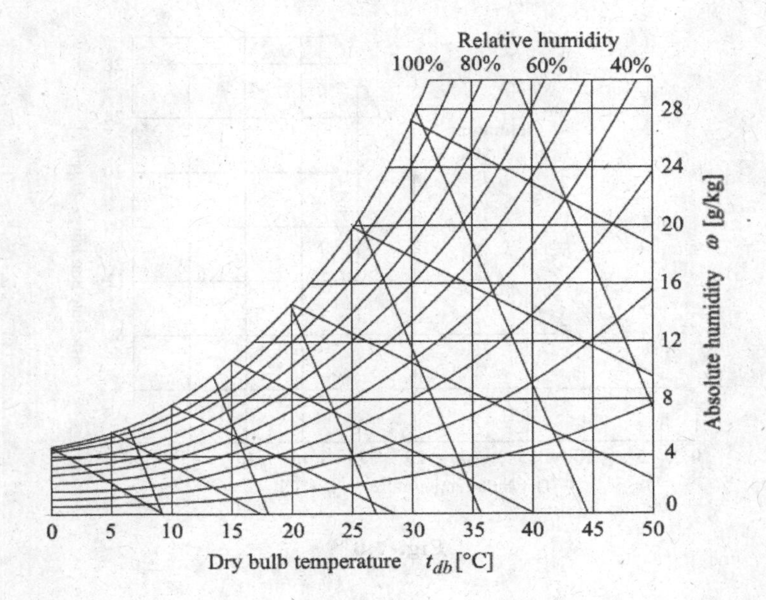

Fig. 7.5

commonly available characteristics — dry-bulb temperature (t_{db}), wet-bulb temperature (t_{wb}) and relative humidity (ϕ) — are known, the others can be read from the chart. Air-conditioning processes, i.e., changes in the condition of the atmosphere, can be represented by the movement of this status point.

Psychrometric charts are created for specific barometric pressures p, but they can be used without significant errors for slightly different pressures. For instance, for locations at not more than 2000 ft (600 m) of altitude it is common practice to use the sea-level psychrometric chart.

Although psychrometric charts may have numerous details, the following lines are important for our study.

The *lines of constant dry-bulb temperature* are vertical, while the *lines of constant absolute humidity* are horizontal, due to the construction of the chart.

The *saturation line* (curve) (Fig. 7.6) limits the area of the diagram to the left. The presence of this limit shows that air at a given temperature and pressure can support only a certain amount of water vapor. The points along this line represent saturation states

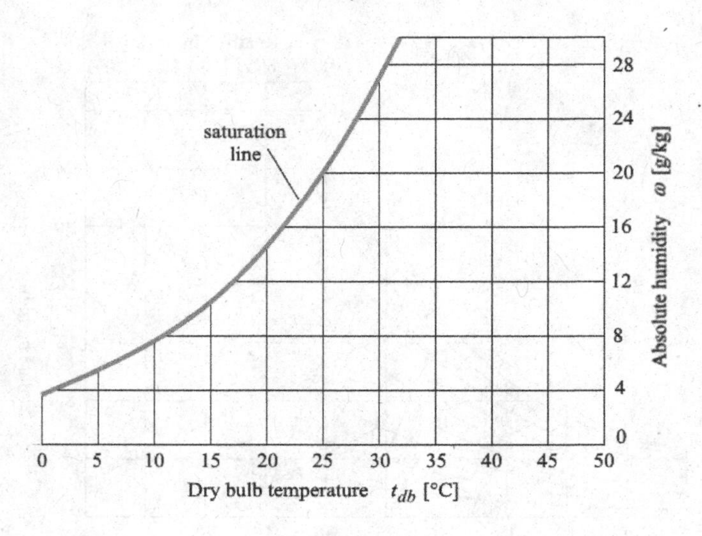

Fig. 7.6

of water vapor. The curve can be plotted using equation (7.15) where $\phi = 100$ % and p_g is taken for various values t_{db}.

The *lines* (curves) *of constant relative humidity* ϕ are curves with positive slope (Fig. 7.7) which decreases with the relative humidity. The line of $\phi = 100$ % is the saturation line. The line corresponding to $\phi = 0$ (dry air) is the horizontal axis. Lines of $\phi = \text{cst}$ are plotted for various temperatures t_{db} with ϕ as a parameter ($\phi = 10$ %, 20 %, etc.), using Eq. (7.15).

The *lines of constant wet-bulb temperature* are parallel lines (Fig. 7.8). They have negative slope and meet the corresponding lines of constant dry-bulb temperature on the saturation line. This is because at saturation $t_{\text{db}} = t_{\text{wb}}$.

The *lines of constant specific volume* are parallel lines with a negative slope, higher than the wet-bulb temperature lines. The specific volume is indicated in Fig. 7.9 in m^3/kg_a.

The *constant enthalpy lines* are parallel lines that almost coincide with the t_{wb} lines (Fig. 7.10). Therefore to avoid the interference and confusion, sometimes the enthalpy lines are not represented inside the chart, instead an enthalpy scale is placed outside the area of the chart, to the left of the saturation line. For moist air condition P,

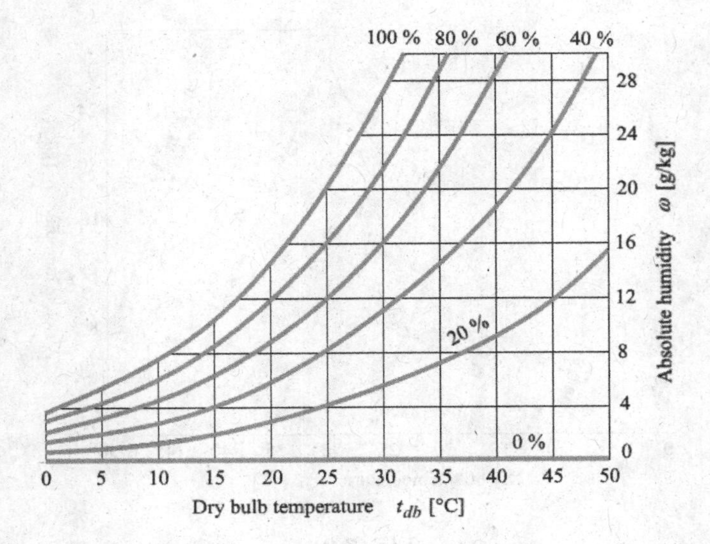

Fig. 7.7

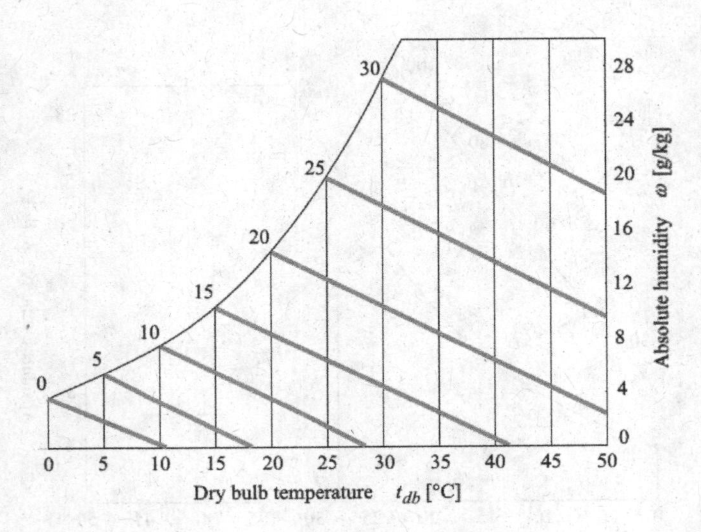

Fig. 7.8

the enthalpy is read at point A, and for point Q, at B. The difference in enthalpy between states P and Q (kJ/kg$_a$) is represented by the segment AB on the enthalpy scale.

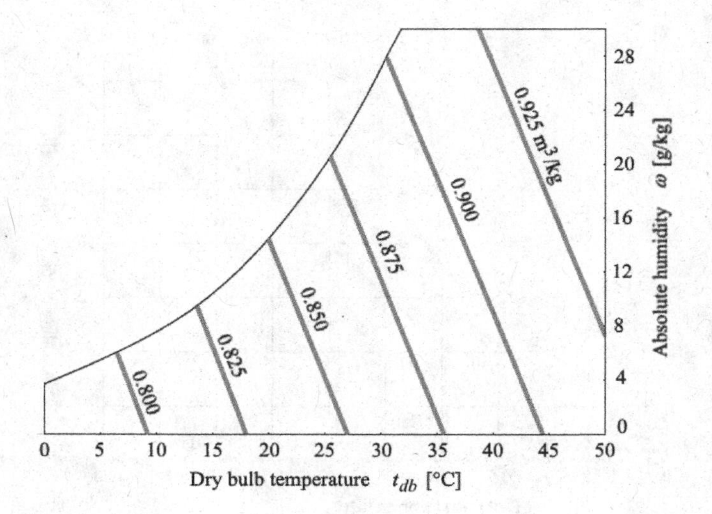

Fig. 7.9

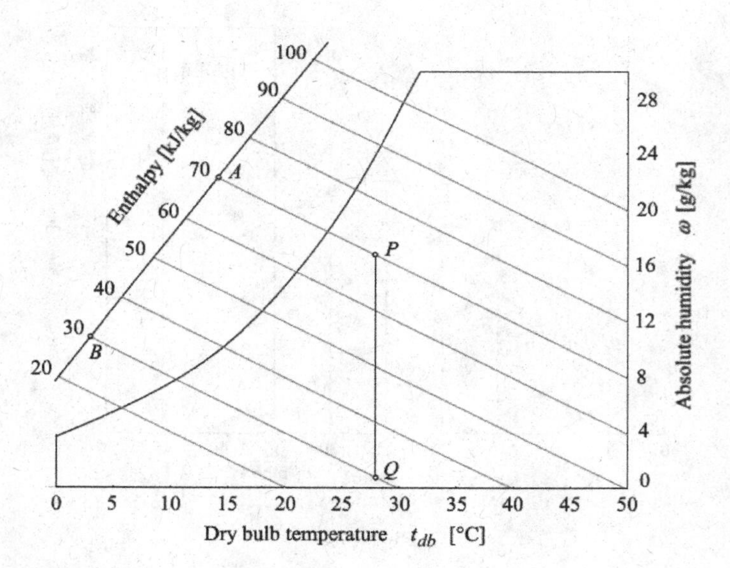

Fig. 7.10

7.2.3. *Air-conditioning processes*

The psychrometric chart is very useful for finding out the properties of moist air (which are required in the field of HVAC) and eliminate a lot of calculations. Some of the common psychrometric processes involved in air-conditioning are as follows:

- *Heating or cooling.* These processes (see Fig. 7.11) refer to the addition or removal of heat, without any change in the moisture content (ω), and result in the change in t_{db}. The status point will move horizontally to the left (cooling) or to the right (heating). Note that while ω does not change, the change in temperature means the relative humidity ϕ changes. It increases if the temperature lowers and vice versa.
- *Dehumidification by cooling.* If, as a result of cooling, the status point moving towards the left ($\omega = \mathrm{cst}$) reaches the saturation line, condensation will start (see Fig. 7.12). The t_{db} corresponding to this point is the *dew-point temperature* of the original atmosphere. If there is further cooling, the status point will move along the saturation line and condensation will occur. This is because at

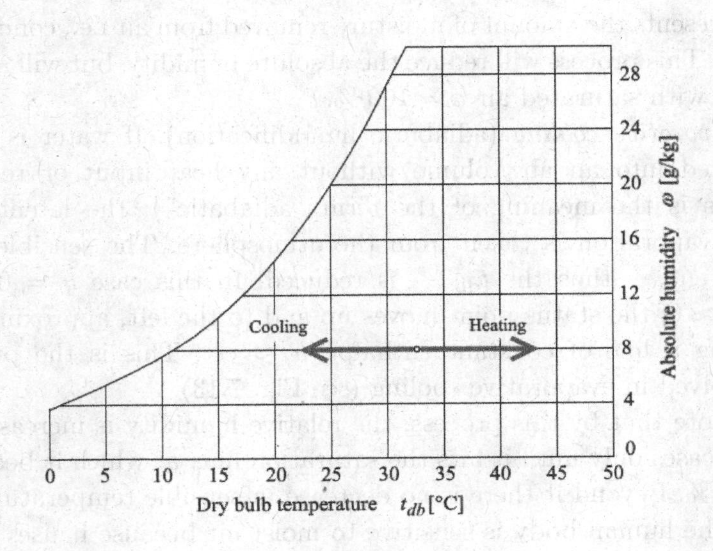

Fig. 7.11

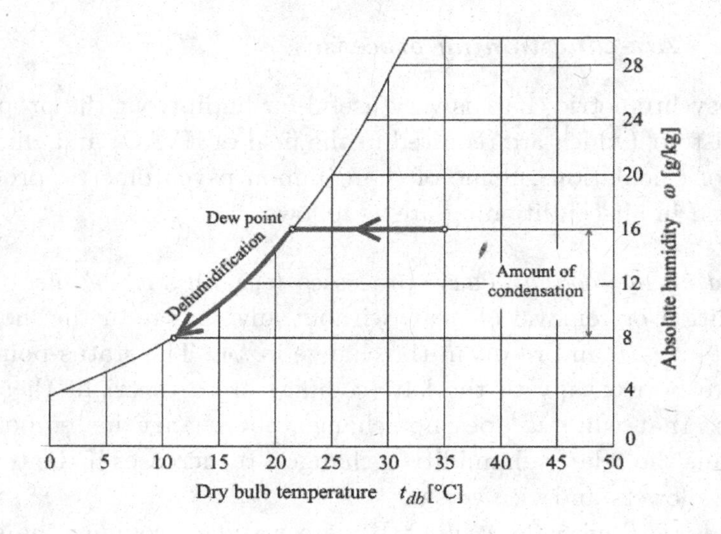

Fig. 7.12

any given temperature atmospheric air can contain only a limited amount of moisture, ω_{sat} (see Sec. 7.2.1). Therefore, no status points can exist to the left of the saturation line.

The reduction in the vertical coordinate (on the ω scale) represents the amount of moisture removed from air i.e., condensed out. This process will reduce the absolute humidity, but will always end with saturated air $\phi = 100\ \%$.

- *Evaporative cooling* (adiabatic humidification). If water is evaporated into an air volume without any heat input or removal (this is the meaning of the term "adiabatic"), the latent heat of evaporation is taken from the atmosphere. The sensible heat content — thus the t_{db} — is reduced. In this case $q = 0$ and, $\Delta h \approx 0$; the status point moves up and to the left, approximately along a line of constant enthalpy $h \approx$ cst. This is the process involved in evaporative cooling (see Fig. 7.13).

Note that by this process, the relative humidity is increased. It increases only until it hits the saturation line, at which it becomes 100 %. Beyond it there is no decrease in sensible temperature.

The human body is sensitive to moist air because it uses evaporative cooling to regulate temperature. Since humans perceive

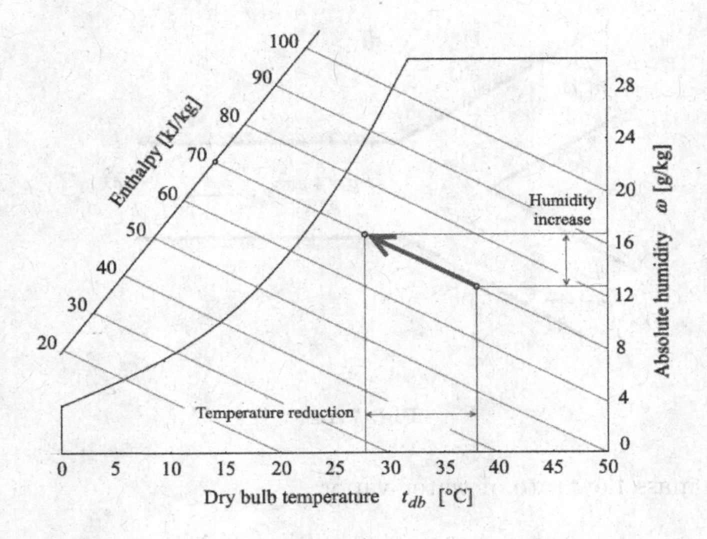

Fig. 7.13

the rate of heat transfer from the body, we feel warmer when the relative humidity is high than when it is low. This is the reason why on a hot and humid day, evaporative cooling is ineffective and uncomfortable.

- *Adiabatic mixing.* Let us consider two air streams 1 and 2 having mass flow rates $\dot{m}_{a1}$ and $\dot{m}_{a2}$, dry-bulb temperatures t_1 and t_2, humidity ratios ω_1 and ω_2 and enthalpies h_1 and h_2, respectively. The adiabatic mixing of the two streams will produce a new condition 3, as shown in Fig. 7.14. Using the laws of conservation of mass and energy one obtains:
- For dry air

$$\dot{m}_{a1} + \dot{m}_{a2} = \dot{m}_{a3}$$

- For water vapor

$$\omega_1 \dot{m}_{a1} + \omega_2 \dot{m}_{a2} = \omega_3 \dot{m}_{a3}$$

The energy balance for the moist air yields:

$$\dot{m}_{a1} h_1 + \dot{m}_{a2} h_2 = \dot{m}_{a3} h_3$$

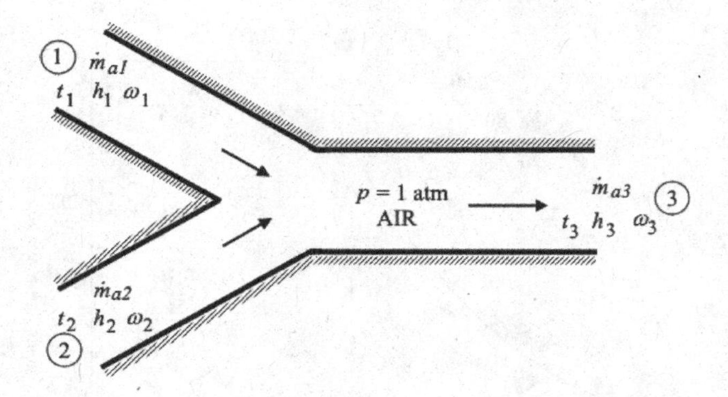

Fig. 7.14

The mass flow rate of water vapor:

$$\dot{m}_{a1}\omega_1 + \dot{m}_{a2}\omega_2 = (\dot{m}_{a1} + \dot{m}_{a2})\omega_3$$
$$\dot{m}_{a1}(\omega_3 - \omega_1) = \dot{m}_{a2}(\omega_2 - \omega_3) \tag{7.20}$$

For the energy balance, a similar expression can be derived:

$$\dot{m}_{a1}(h_3 - h_1) = \dot{m}_{a2}(h_2 - h_3) \tag{7.21}$$

Therefore:

$$\frac{\dot{m}_{a1}}{\dot{m}_{a2}} = \frac{\omega_2 - \omega_3}{\omega_3 - \omega_1} = \frac{h_2 - h_3}{h_3 - h_1} \tag{7.22}$$

On the psychrometric chart, this is the equation of a line linking the points 1 and 2. Point 3, representing the final state of the mixture, lies on the straight line 1–2 and divides the line in the inverse ratio of the mixing mass flow rates (Fig. 7.15). Therefore:

$$\frac{\dot{m}_{a1}}{\dot{m}_{a2}} = \frac{h_2 - h_3}{h_3 - h_1} = \frac{\omega_2 - \omega_3}{\omega_3 - \omega_1} = \frac{t_2 - t_3}{t_3 - t_1} = \frac{\overline{23}}{\overline{13}} \tag{7.23}$$

Equation (7.23) shows that the state of the mixture can be obtained from a simple geometric construction, using the lever rule; the properties of the mixture can be read directly from the chart.

Many industrial processes, including those occurring in steam power plants, require large amounts of cooling water. When natural

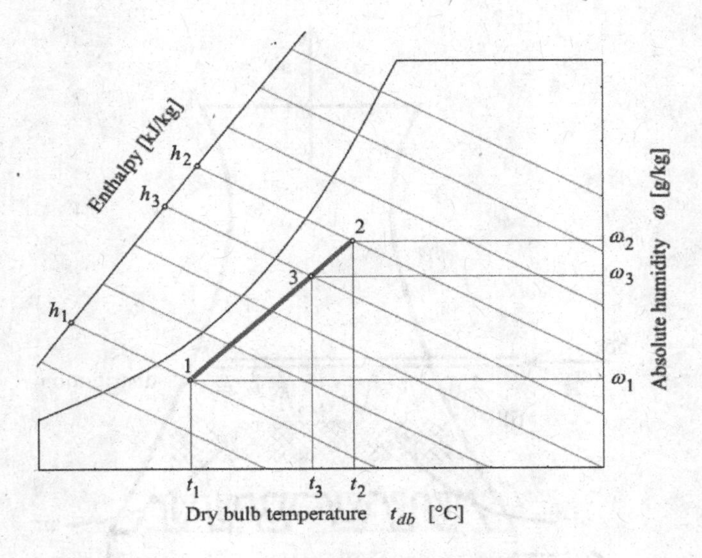

Fig. 7.15

sources are not available, water needs to be cooled after use and then recirculated. An effective method is to use the principle of evaporative cooling in a **cooling tower**. Warm water from an industrial process is pumped to the cooling tower through pipes. Then water is sprayed through nozzles onto banks of material called *fill*, thus exposing as much water surface area as possible for maximum air–water contact. As the water flows through the cooling tower, it is exposed to air, which is being pulled through the tower by natural draft (as shown in Fig. 7.16) or by an electric motor-driven fan. When the water and air are in direct contact, a small amount of water is evaporated, creating a cooling action. The cooled water is then pumped back to the condenser or process equipment where it absorbs heat, and the cycle continues. The cooling effect can reduce the water temperature to within only a few degrees of the dry-bulb temperature. The disadvantage is that some treated water (called *makeup water*) is still needed to cover continuously the evaporation losses.

Air-conditioning plants represent another category of devices where thermodynamic processes involving moist air are utilized.

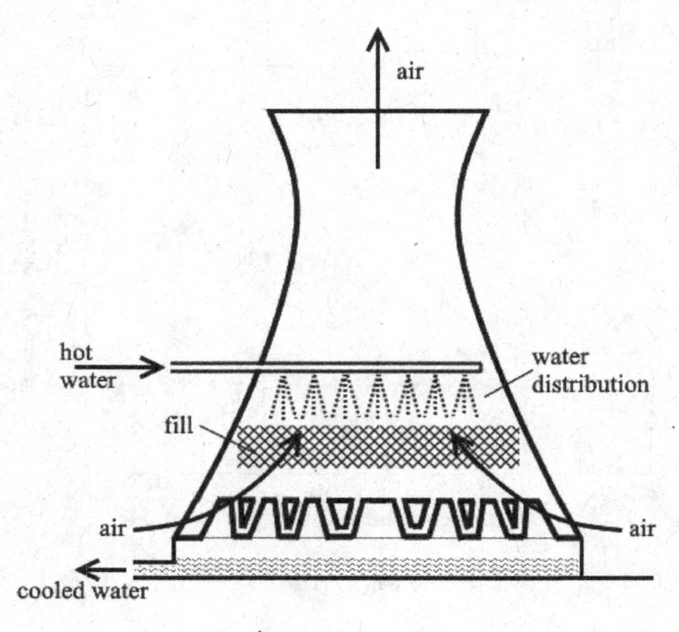

Fig. 7.16

These plants are capable of continuous supply of air at some regulated temperature and relative humidity. The levels of temperature and humidity are imposed either by the human comfort or by the requirements of some industrial equipment for correct operation.

Properties of gas mixtures: moist air — problems

7.1. A mixture of gases consists of two components of equal mass. Their densities, at a certain pressure and temperature, are $\rho_1 = 1.1$ kg/m^3 and $\rho_2 = 0.8$ kg/m^3, respectively. What is the density of the mixture under the same conditions?

7.2. A mixture of gases consists of nitrogen and water vapor. The quantity of nitrogen is such that it occupies a volume of 0.79 m^3 at $p_{N_2} = 1$ atm and $t_{N_2} = 0$ °C, while the mass of water vapor is 0.09 kg. Determine: (a) the mass fraction of each constituent; (b) the specific gas constant for the mixture; (c) the partial pressures of the constituents when the mixture occupies a volume $V = 22$ dm^3 at $t = 25$ °C.

7.3. The absolute humidity (or moisture content) ω of moist air represents: (a) the quantity of water vapor in 1 kg of moist air; (b) the quantity of water vapor corresponding to 1 kg of dry air; (c) the quantity of water vapor in a given volume.

7.4. Vapor turning into liquid will: (a) absorb from the ambient the latent heat of condensation; (b) release to the ambient the latent heat of condensation; (c) have constant temperature during condensation.

7.5. Heating of humid air at constant absolute humidity, ω, results in: (a) increase of its enthalpy and relative humidity; (b) increase of its temperature and decrease of its relative humidity; (c) increase of its temperature and entropy.

7.6. A room contains moist air at $p = 101$ kN/m^2. The partial pressure of water vapor is $p_v = 51$ mbar. Determine: the partial pressure of dry air and the absolute humidity.

7.7. Humid air having $t_1 = 29$ °C and $\phi_1 = 20$ % is cooled (without any change in the moisture content ω) to $t_2 = 14$ °C. Determine the absolute humidity and the final relative humidity ϕ_2.

7.8. Thermal measurements performed for a dwelling show that the windows' temperature on a winter day is 5 °C while the air temperature in the room is 23 °C and the barometric pressure is 748 mmHg. Determine the maximum relative humidity that can be accepted in the room without condensation on windows. Also, determine the corresponding partial pressures of air and water vapor, the absolute humidity and the density of moist air.

7.9. In an air conditioning plant 65 m^3/min of moist air is passed through a heating chamber. The air is introduced at $t_1 = 13$ °C and $\phi_1 = 80$ % and exits at 24 °C. Considering that the process occurs at constant atmospheric pressure, determine the final relative humidity ϕ_2 and the rate of heat transfer to the air.

7.10. A stream of 8 kg/min outdoor air at 5 °C and 20 % of relative humidity is adiabatically mixed with 20 kg/min of recirculated air at 30 °C and 80 % relative humidity. Find the temperature and the absolute and relative humidities of the resulting mixture.

Chapter 8

Vapor Power Cycles

Cycles during which work is done by the system using the heat received from a hot reservoir are called **power cycles** (see also Sec. 6.1).

Power cycles are often classified by the type of working fluid into:

- vapor cycles
- gas cycles

Regardless of the type, these are ideal cycles, in the sense that they represent prototypes of actual cycles and their efficiencies approach but are always less than the Carnot cycle efficiency. In practice, actual cycles differ from ideal because of inherent irreversibilities and for other practical reasons. However, the study of ideal cycles leads to conclusions that are applicable to real cycles.

Vapor power cycles are used to study the operation of steam power plants. During these cycles, the working fluid undergoes phase changes, from liquid to vapor and back to liquid.

The study of vapor power cycles makes use of p–v, T–s (Appendix E, Fig. A.1) and h–s (Appendix E, Fig. A.2) diagrams for water. Relevant property tables are: properties of saturated water (liquid–vapor) (Appendix E, Tables A.2 and A.3), and properties of superheated water vapor (Appendix E, Table A.4). Similar information can be found in [11], Tables A.4–A.7.

The Carnot cycle is the most efficient cycle operating between two specified temperature limits, T_H and T_C. A *Carnot vapor cycle*

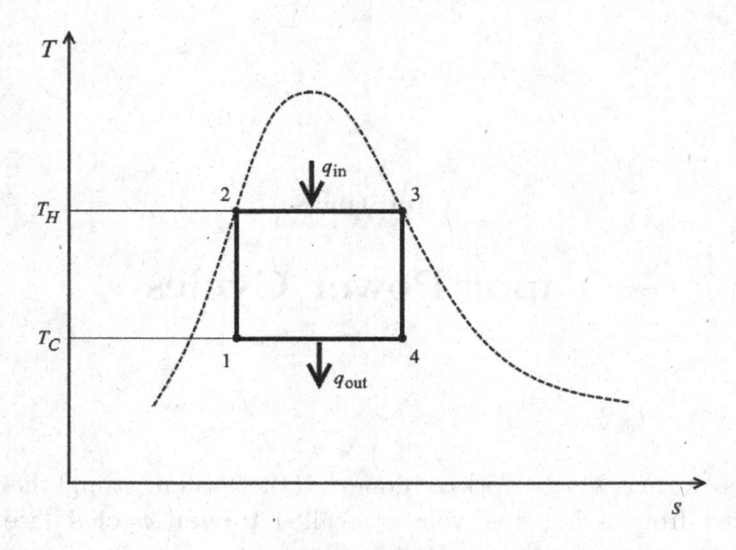

Fig. 8.1

is represented in Fig. 8.1. It consists of two isothermal processes (vaporization 2–3 and condensation 4–1) and two isentropic processes (compression 1–2 and expansion 3–4). However, such a cycle cannot be used for the operation of a power plant because it is affected by some impracticalities: the limitation of the heating process to a two-phase system (points 2 and 3); the expansion of low-quality steam in a turbine, leading to severe erosion of the blades from water droplets (process 3–4); isentropic compression of a two-phase system (process 1–2). Most of these impracticalities are eliminated if the heating process does not stop at the saturation line, but continues in the superheated steam region, and if the condensation process is extended to the saturated liquid line. This results in the creation of a new cycle, called the **Rankine cycle**.[a]

8.1. The Rankine cycle

The Rankine cycle is the ideal cycle for vapor power plants. It does not involve any internal irreversibilities and consists of the following

[a]Named after William John Macquorn Rankine (1820–1872), a Scottish civil engineer, physicist and mathematician, who described it in 1859.

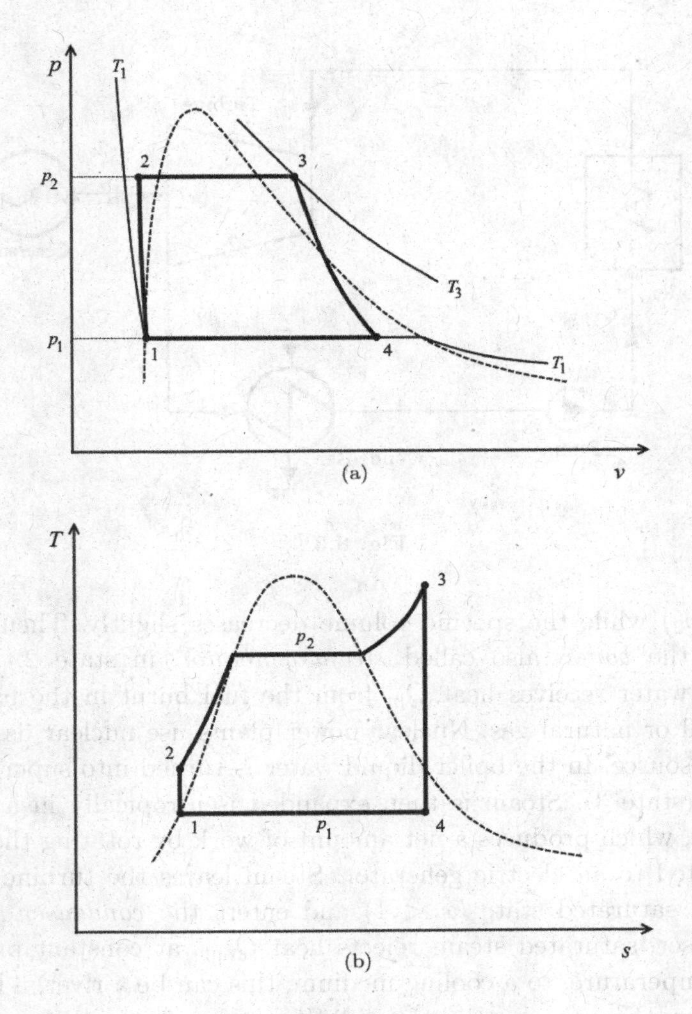

Fig. 8.2

processes (see Fig. 8.2):

1–2: isentropic compression of water in a *pump*;
2–3: constant pressure heat addition in a *boiler*;
3–4: isentropic expansion in a *turbine*;
4–1: constant pressure heat rejection in a *condenser*.

Saturated water in state 1 is isentropically compressed in the boiler *feed pump*. Pressure increases from condenser level (p_1) to the boiler

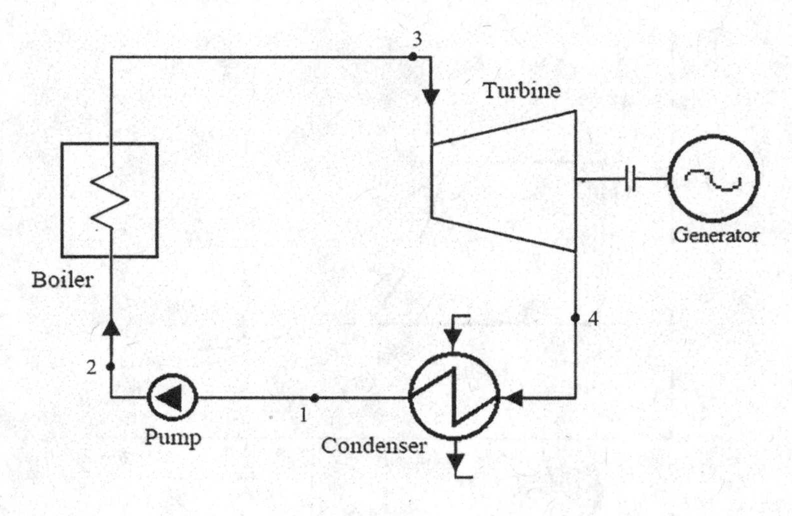

Fig. 8.3

level (p_2) while the specific volume decreases slightly. Then water enters the *boiler*, also called *steam generator*, in state 2. In the boiler, water receives heat Q_{in} from the fuel burnt in the furnace: coal, oil or natural gas. Nuclear power plants use nuclear fission as a heat source. In the boiler, liquid water is turned into superheated steam (state 3). Steam is then expanded isentropically in a steam *turbine*, which produces a net amount of work by rotating the shaft connected to an electric generator. Steam leaves the turbine (state 4) in a saturated state ($x < 1$) and enters the *condenser*. In the condenser, saturated steam rejects heat Q_{out}, at constant pressure and temperature, to a cooling medium; this can be a river, a lake or the atmosphere (in a cooling tower). Steam then leaves the condenser as saturated water (state 1), and this completes the cycle (see also Fig. 8.3).

In a $T{-}s$ diagram (Fig. 8.2(b), not to scale), the area under the process line represents the heat exchanged; therefore:

- Area under 2–3: q_{in}
- Area under 4–1: q_{out}
- $q_{\text{in}} - |q_{\text{out}}| = w_{\text{net}}$

8.2. Energy analysis of the Rankine cycle

Considering the four components of the Rankine cycle as steady-flow devices as presented in Sec. 5.2, for each process the steady-flow energy equation for one unit mass can be written as

$$q - w = \Delta h = h_{\text{final}} - h_{\text{initial}}$$

This general equation can be expressed in specific ways for each device, as follows:

Pump, process 1–2 ($q = 0$): $w_{\text{pump,in}} = h_1 - h_2$; since $dh = T\,ds + vdp$, it follows that, considering liquid water as practically incompressible, $v_1 = v_{f1} \approx v_2$,

$$w_{\text{pump,in}} \approx v_{f1}(p_1 - p_2) \tag{8.1}$$

Boiler, process 2–3 ($w = 0$):

$$q_{\text{in}} = h_3 - h_2 \tag{8.2}$$

Turbine, process 3–4 ($q = 0$):

$$w_{\text{turb,out}} = h_3 - h_4 \tag{8.3}$$

Condenser, process 4–1 ($w = 0$):

$$q_{\text{out}} = h_1 - h_4 \tag{8.4}$$

The thermal efficiency:

$$\eta_{\text{th}} = \frac{w_{\text{net}}}{q_{\text{in}}} = 1 - \frac{|q_{\text{out}}|}{q_{\text{in}}}$$

$$\eta_{\text{th}} = \frac{w_{\text{net}}}{q_{\text{in}}} = \frac{w_{\text{turb,out}} - |w_{\text{pump,in}}|}{q_{\text{in}}} = \frac{(h_3 - h_4) - |h_1 - h_2|}{h_3 - h_2} \tag{8.5a}$$

Because the specific volume of liquid is thousands of times smaller than the average specific volume of steam, the *feed pump term* $|h_1 - h_2|$ is very small compared to the total enthalpy drop in the turbine[b]; on a $T\!-\!s$ diagram drawn to scale, points 1 and 2 almost

[b]For gas-turbine power plants, the ratio $r_{\text{bw}} = |w_{\text{in}}|/w_{\text{out}}$ is called **back work ratio**. For a steam-turbine power plant r_{bw} is very small, typically 1 % or 2 %.

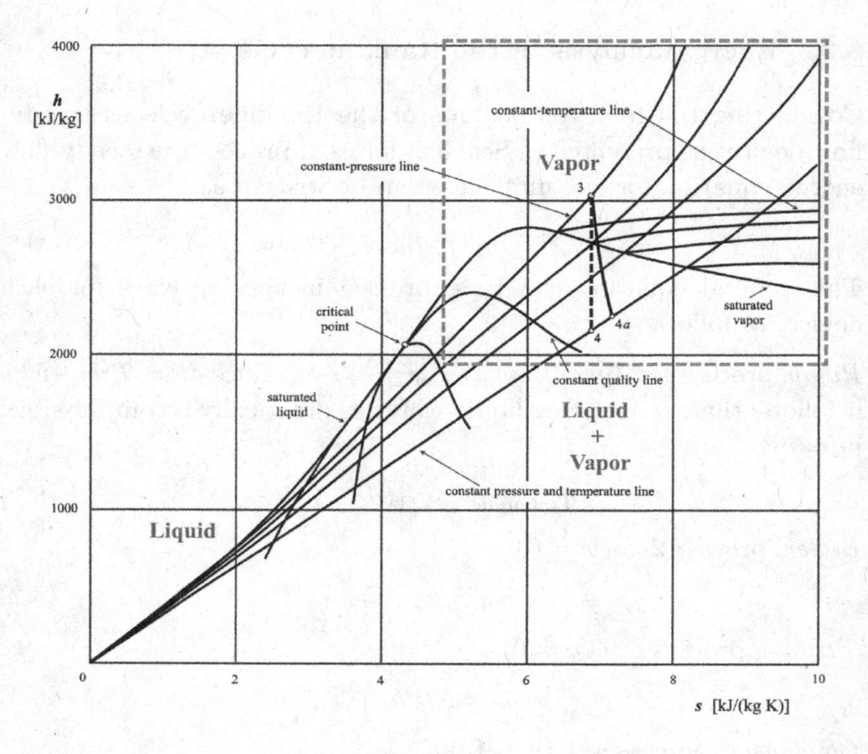

Fig. 8.4

coincide. Therefore

$$\eta_{\text{th}} \approx \frac{h_3 - h_4}{h_3 - h_2} \tag{8.5b}$$

For the steam work, one commonly used diagram is the $h-s$ diagram (see Fig. 8.4), known as the **Mollier diagram** for water, or *Mollier chart*, named after Dr. Richard Mollier of Dresden who first devised the idea of such a diagram in 1904. On the diagram the lines of constant pressure and constant temperature are divergent from the origin and in the saturation region they coincide. The saturation curve for liquid and steam is also represented, as on the $T-s$ diagram, but here the critical point is not at the peak point of the curve, but on the left side. The density of lines is quite high on the bottom left corner, but a very good resolution can be obtained in the superheated and saturated steam and this portion of the diagram is used in practice (see Fig. 8.4, the dashed rectangle).

The advantage of the $h-s$ diagram is that the isentropic expansion in the turbine can be conveniently plotted as a vertical line, the amount of work (isentropic or not) being easily determined as the distance between points 3 and 4 (or 4a) on the enthalpy scale. Also, the steam quality can be directly read from the diagram. This eliminates the need for calculations using tables for superheated and saturated steam.

The actual vapor cycle differs from the ideal Rankine cycle as a result of irreversibilities affecting its processes: fluid friction and heat loss to the surroundings. Of a particular importance are irreversibilities affecting the pump and the turbine (see diagram in Fig. 8.4). While ideally the flow through these devices would be isentropic, the actual processes are not: the pump consumes more work than in the ideal case, while the turbine produces less work. The deviations from the isentropic processes can be accounted for by using **isentropic efficiencies**, defined for the pump and for the turbine as follows (see Fig. 8.5):

$$\eta_P = \frac{w_{Ps}}{w_{Pa}} = \frac{h_1 - h_2}{h_1 - h_{2a}} \tag{8.6}$$

$$\eta_T = \frac{w_{Ta}}{w_{Ts}} = \frac{h_3 - h_{4a}}{h_3 - h_4} \tag{8.7}$$

where indices s and a refer to the isentropic and actual situations, respectively.

The overall efficiency of the cycle becomes

$$\eta = \eta_{\text{th}} \eta_P \, \eta_T \tag{8.8}$$

Due to the low impact of the feed pump term, η_P is often neglected. A typical value for η_T is 85 %. Modern boilers of large capacity used in power plants have an efficiency ranging from 80 to 90 %.

For a realistic analysis of the operation of a steam power plant, other factors need to be considered also: subcooling of liquid in the condenser, pressure losses in all components, fluid leaks, air leaks into the condenser, power consumed by auxiliary equipment, etc.

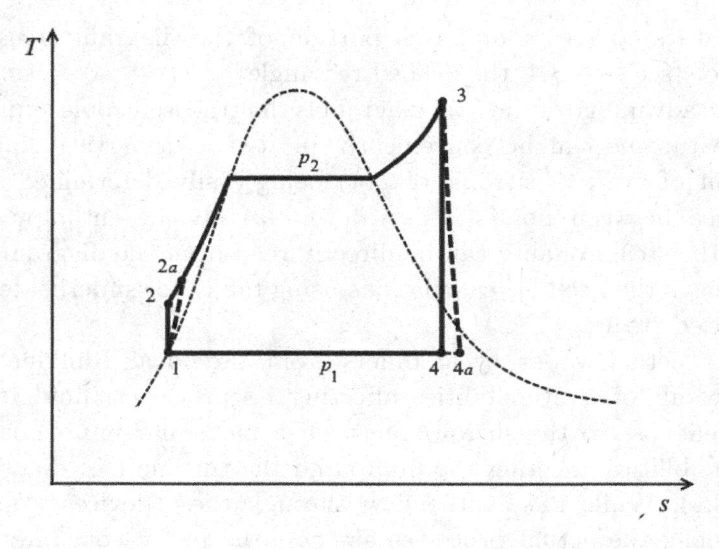

Fig. 8.5

8.3. Means to increase the efficiency of the Rankine cycle

As in the case of a Carnot cycle, the wider the range of temperature, the more efficient becomes the cycle. For the Rankine cycle, this can be accomplished in two ways:

- by increasing the average temperature at which the working fluid receives heat in the boiler, and
- by decreasing the temperature at which heat is rejected from the working fluid in the condenser.

Lowering the condenser pressure. As one can see from the $T-s$ diagram in Fig. 8.6, by lowering the condenser pressure, the lower temperature of the cycle decreases and the net work produced by the cycle increases. However, the lowest possible temperature of the condensing steam is dictated by two factors:

- the temperature of the natural sink of heat (atmosphere, lake, river, ocean) which, for areas with temperate climate is around 15 °C (annual average), and

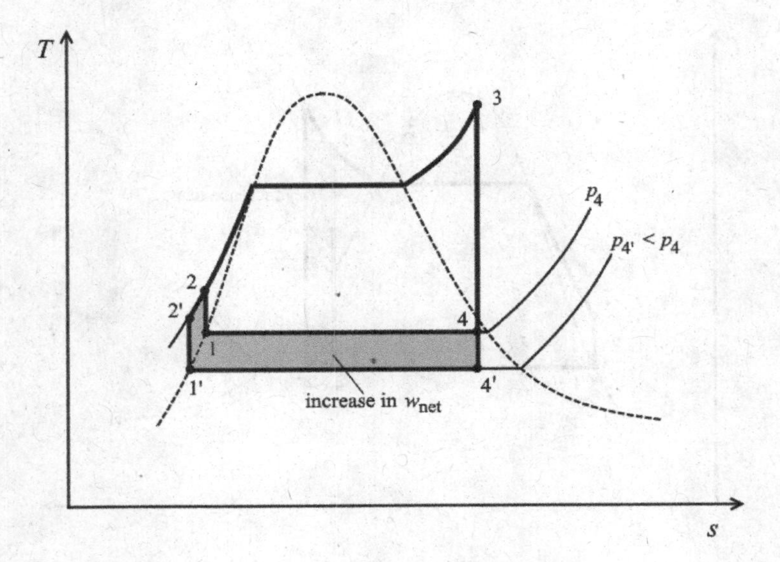

Fig. 8.6

- the temperature difference between fluids required for the heat transfer in the condenser which, for a reasonable size of the heat exchanger, has to be of 10–15 °C.

Therefore, the lowest practical condensing temperature in the cycle is 25–30 °C which corresponds to a pressure (see Appendix E, Table A.2) of about 0.032–0.042 bar.

Superheating the steam to higher temperatures will increase the average temperature at which the working fluid receives heat in the cycle (see the $T-s$ diagram in Fig. 8.7). This will result in an increase of the cycle work. The cycle efficiency increases continuously with temperature. The only limiting factor is the strength of the materials available. The metallurgical limit for long-life plant is about 600–650 °C at the present time.

Increasing the boiler pressure. This will increase the saturation temperature and, implicitly, the average temperature at which the working fluid receives heat (see Fig. 8.8). Operating pressures of boilers can be brought to super-critical values, over 300 bar (30 MPa);

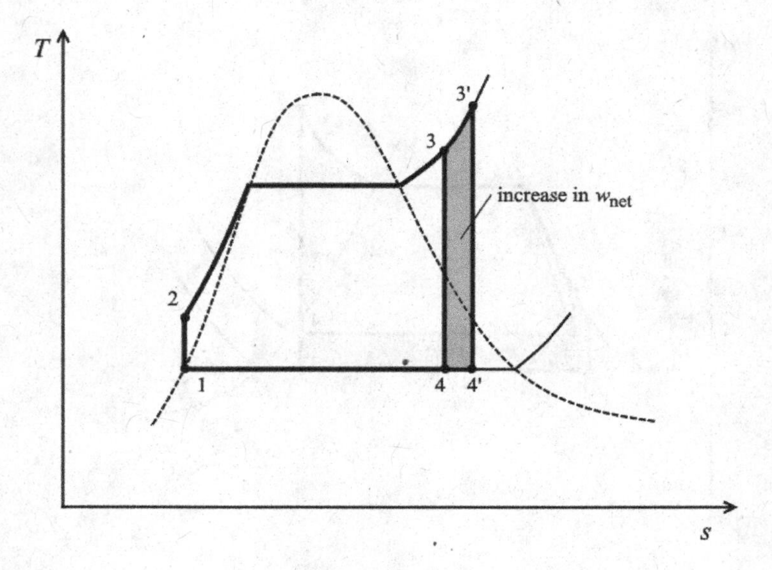
T
3'
3
increase in w_{net}
2
1
4
4'
s

Fig. 8.7

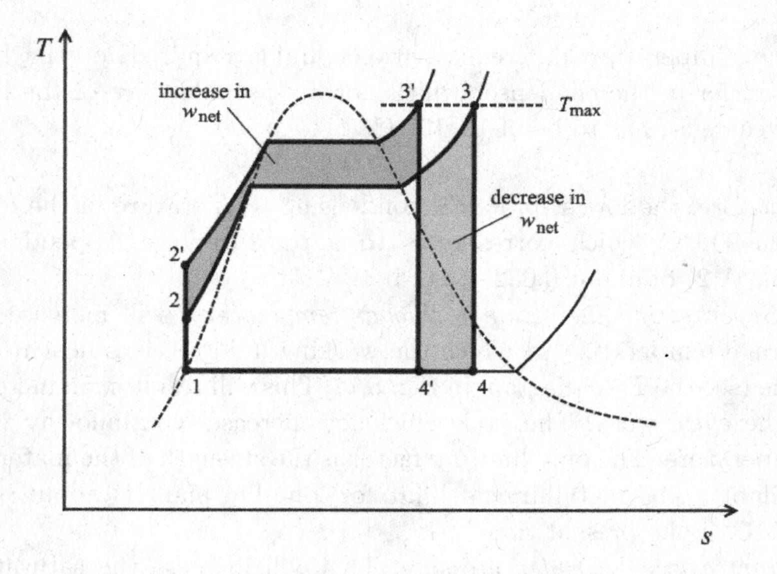
T
increase in w_{net}
3'
3
T_{max}
decrease in w_{net}
2'
2
1
4'
4
s

Fig. 8.8

this is the case of the Benson-type boilers,[c] which can be installed in steam plants with net power outputs exceeding 1000 MW. However, the $T-s$ diagram shows that for a fixed turbine inlet temperature $T_{\max}$, with higher boiler pressures the cycle shifts to the left and the turbine exhaust becomes wetter. In practice, the steam quality at the turbine exit is not allowed to fall below about 0.88 [1, p. 218] because the water droplets in the steam erode the blading and reduce the turbine isentropic efficiency. This drawback of using high boiler pressure can be corrected by reheating the steam.

With the reheat cycle, superheated steam expands first in a *high-pressure turbine* to some intermediate pressure (process 3–4, Fig. 8.9) and then is sent back to the boiler for *reheat* at constant pressure (process 4–5). Reheated steam, usually at the original inlet temperature ($T_5 = T_3$), expands in a *low-pressure turbine* to the condenser pressure (process 5–6).

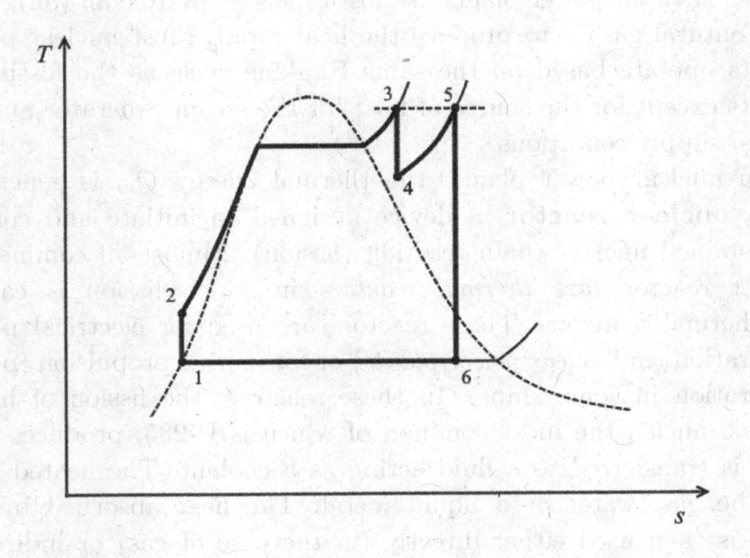

Fig. 8.9

[c]Named after Mark Benson (born Mark Müller), a German engineer, the inventor of a supercritical boiler.

Reheating makes only a little difference to the ideal cycle efficiency [1, p. 221], moreover, it can increase or decrease the efficiency when compared to the Rankine cycle having the same conditions in the boiler and condenser, depending on the point where the expansion is interrupted for reheating. The most significant positive impact comes from the increased turbine exhaust steam quality and the reduction of the specific steam consumption (the amount of steam necessary to produce one unit of energy at the turbine).

Other means of increasing the efficiency of a Rankine cycle include *regenerative feedwater heating* (or simply regeneration) [7, p. 202], cogeneration [11, p. 592] and combining gas and vapor power cycles ([11, p. 597; 34]).

8.4. Steam cycles for nuclear power plant

Classical vapor power plants use fossil fuels — petroleum (oil), coal, and natural gas — to produce the heat input. Most nuclear power plants operate based on the same Rankine cycle as the fossil fuel plants except for the source of heat for the steam generator and for steam supply conditions.

In nuclear power plants, the thermal energy Q_{in} is generated by a **nuclear reactor**, a device designed to initiate and control a sustained nuclear chain reaction (fission). Almost all commercial power reactors are *thermal reactors* in which fission is caused by thermal neutrons. These reactors are used for electrical power generation (in *nuclear power plants*) or for marine propulsion (power generation in some ships). In these reactors, the fission of heavy atomic nuclei, the most common of which is U-235, produces heat that is transferred to a fluid acting as a coolant. The heated fluid can be gas, water or a liquid metal. The heat absorbed by the fluid is then used either directly (in the case of gas) or indirectly (in the case of water and liquid metals) to generate steam. The heated gas or the steam is then fed into a turbine driving an alternator.

The most common thermal reactors are those using plain distilled water (light water) as a coolant. Depending on the thermodynamic conditions of the water used to cool uranium fuel elements in the reactor vessel, we distinguish between:

- boiling water reactors (BWR) — operating at a pressure that allows boiling of the coolant water adjacent to the fuel elements.
- pressurized water reactors (PWR) — where water is at about the same temperature as in the BWR but at a higher pressure, so that the reactor coolant remains a liquid throughout the reactor coolant loop.

The BWR's distinguishing feature is that the reactor vessel serves as the boiler for the nuclear steam supply system (Fig. 8.10). Saturated steam at pressures around 70 bar is generated in the reactor vessel by the controlled fission of enriched uranium fuel and goes directly to the turbo-generator to produce electricity [35]. The operating efficiency of BWR plants is around 32 % [36].

The PWR (also known as VVER if of Russian design) is second generation nuclear power reactor that uses ordinary water under high pressure as coolant for the reactor core and as the moderator to control the chain reaction. This highly pressurized hot water then exchanges heat with a lower pressure system, where water turns to

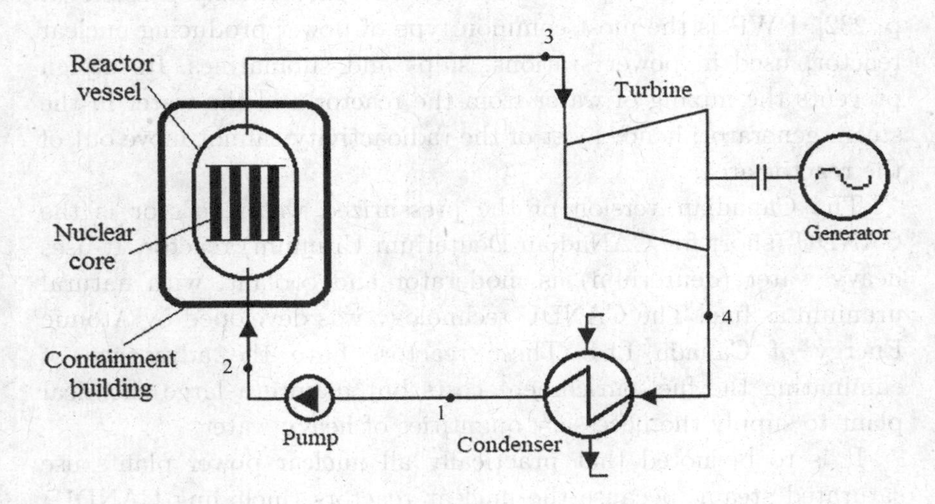

Fig. 8.10

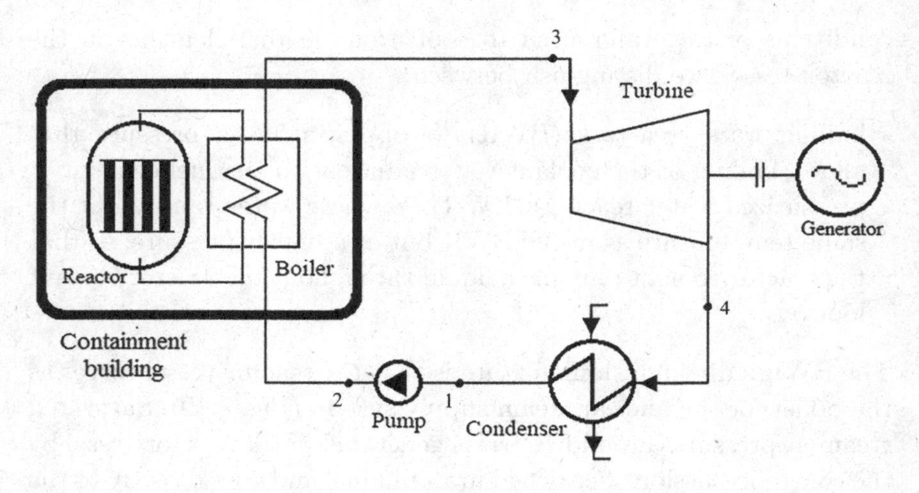

Fig. 8.11

steam and drives the turbine (Fig. 8.11). Currently PWR operate at about 160 bar and 315 °C in the primary circuit, and about 60 bar (275 °C) in the lower-pressure secondary circuit. This provides a higher thermal efficiency than the BWR, but the reactor is more complicated and costlier to construct. Under particular conditions, a certain amount of superheating becomes theoretically possible [1, p. 232]. PWR is the most common type of power producing nuclear reactor, used in power stations, ships and submarines. Its design prevents the mixing of water from the reactor and the water in the steam generator; hence most of the radioactivity cannot move out of the reactor area.

The Canadian version of the pressurized water reactor is the CANDU (short for CANadian Deuterium Uranium) reactor. It uses heavy water (deuterium) as moderator and coolant, with natural uranium as fuel. The CANDU technology was developed by Atomic Energy of Canada Ltd. These reactors have the advantage of eliminating the fuel enrichment costs but require a large chemical plant to supply the necessary quantities of heavy water.

It is to be noted that practically all nuclear power plants use saturated steam, because the nuclear reactors (including CANDU) are not suitable for superheating and superheating using auxiliary fuel is very difficult to manage (two different circuits need to be adjusted simultaneously). The subsequent problem of steam quality is solved using *moisture separators* or "steam dryers" to separate water droplets from steam. These dryers are used as the final stage of water separation within a reactor vessel (BWR plants) or within the steam generator vessel (PWR plants) and between the stages of the turbine, to prevent rapid blade wear from water erosion.

Vapor power cycles — problems

8.1. In some power plant 50 t/h of steam is generated at 60 bar and 400 °C then passed through a throttling valve which reduces the pressure to 50 bar. At that point steam is introduced in a turbine where it expands adiabatically to a pressure of 2 bar. The isentropic efficiency of the turbine is 86 %. Represent the processes in the $h{-}s$ diagram and determine: (a) the temperature of steam at turbine inlet; (b) the steam quality at the turbine exit; (c) the power produced by the turbine.

8.2. A steam power plant operates based on the Rankine cycle with superheat. The steam parameters at the turbine inlet are $p_3 = 100$ bar and $t_3 = 500$ °C. The condenser pressure is $p_4 = 0.05$ bar. Represent the cycle in a $T{-}s$ diagram and determine, for all representative points of the cycle, the values of state quantities p, t, h, s, as well as the steam quality at turbine exit, x.

8.3. A steam power plant operates on a Rankine cycle with reheat as shown in Fig. 8.12. Steam leaves the boiler at 20 MPa and 700 °C. The high-pressure turbine exhausts to 0.4 MPa and the steam is then reheated in the boiler to 600 °C. The low-pressure turbine exhausts to 0.004 MPa. Represent the cycle in a $T{-}s$ diagram and determine: (a) the thermal efficiency of the plant; (b) the mass flow rate of steam leaving the boiler required to produce 50 MW of power.

8.4. A steam power plant operates based on an ideal Rankine cycle. Superheated vapor enters the turbine at 8 MPa, 500 °C. The condenser pressure is 6 kPa. The net power of the cycle is

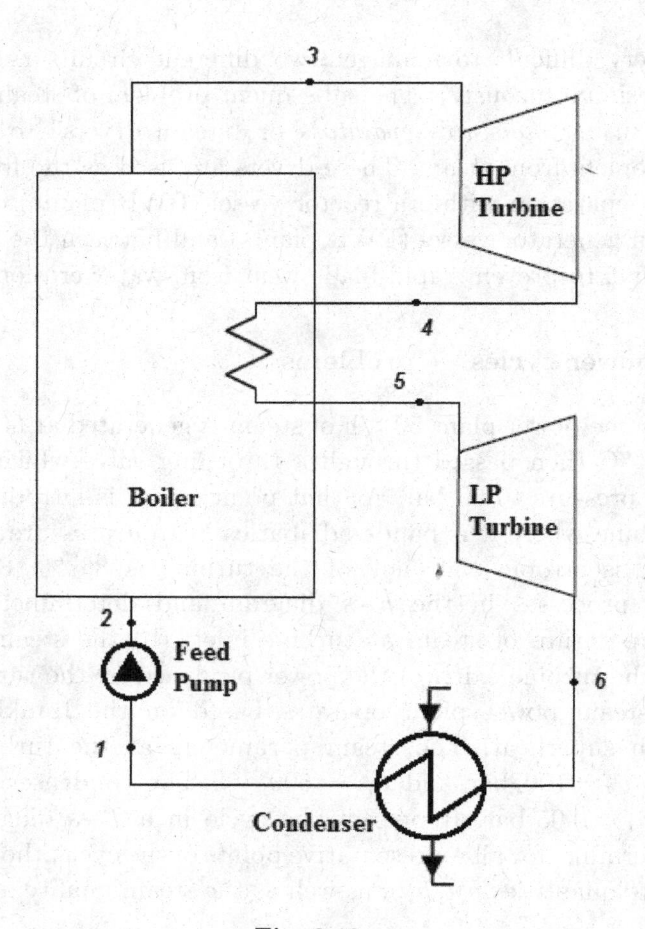

Fig. 8.12

80 MW. Represent the cycle in a $T-s$ diagram and determine: (a) the rate of heat received by the working fluid in the boiler; (b) the thermal efficiency of the cycle; (c) the mass flow rate of condenser cooling water if the cooling water enters the condenser at 17 °C and exits at 38 °C. Any pressure losses are negligible.

Chapter 9

Gas Power Cycles

9.1. Basic considerations

Gas power cycles are essentially used to study the operation of internal-combustion[a] (IC) engines and power plants equipped with gas-turbines. The working fluid is air, or a mixture of air and combustion products, and will be considered a perfect gas. The processes in gas-turbine plants are steady-flow processes carried out in separate components. The processes occurring in an IC engine are non-flow processes taking place in a cylinder fitted with a piston. The cycles encountered in actual devices are complex and difficult to analyze because of the presence of irreversibilities and the fact that the parameters of the working fluid vary continuously. In order to make an analytical study of a cycle feasible, some idealizations are necessary. Only the ideal cycles will be discussed in this section.

The Carnot cycle is used as reference for gas power cycles, too, since it has the highest thermal efficiency of all cycles working between the same temperature limits.

9.2. Reciprocating internal-combustion engines

There are two types of IC engines: *reciprocating* and *rotary* engines. A reciprocating IC engine is a heat engine where the chemical energy

[a]The term "internal combustion" is used to show that the fuel is burnt within the boundaries of the system.

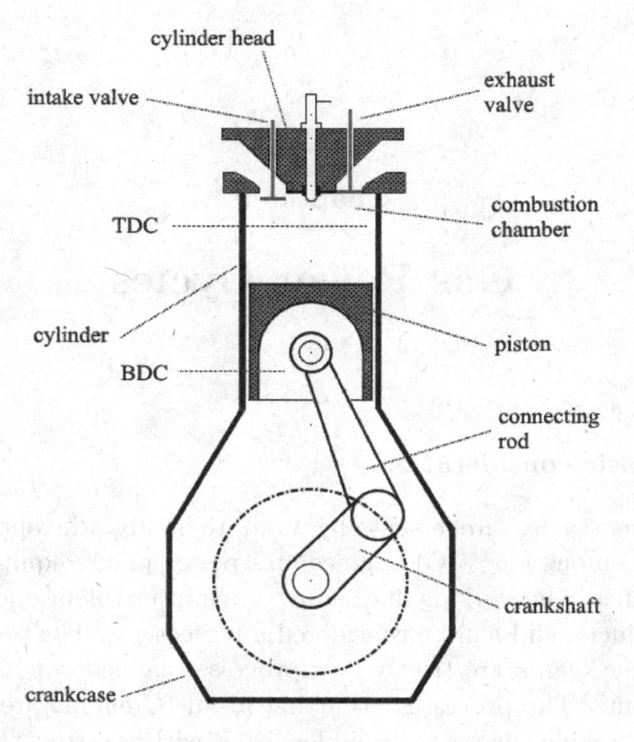

Fig. 9.1

in the fuel is converted into work in a chamber having a mobile wall: a piston which moves back and forth inside a cylinder. This is the most widely used type of engine. In the case of rotary engines, a rotor rotates inside the engine to produce power. The key components of a reciprocating IC engine, also often known as a *piston engine*, are presented in Fig. 9.1. A reciprocating engine, uses one or more cylinders with pistons that move back and forth inside, between two extreme positions called *dead centers*. The *top dead center* (TDC) is the position of a piston in which it is farthest from the crankshaft. The *bottom dead center* (BDC) is the position nearest to the crankshaft. All reciprocating engines have some common features: air and some atomized fuel are introduced inside a cylinder in the space between the *piston* and the *cylinder head* called *combustion chamber*, and the flammable mixture is ignited. The hot combustion gases expand, pushing the piston to the BDC. The piston is returned to the cylinder top (TDC) either by a flywheel or the power from other pistons connected to the same shaft, while the expanded gases are evacuated from the cylinder by this stroke. The linear movement of the piston is converted to a rotating movement via a *connecting rod* and a *crankshaft.* The distance traveled by the piston between the TDC and the BDC is called *stroke.* The volume swept by the piston inside the cylinder in a single stroke is called *displacement volume* or *swept volume* (V_s) and the volume of the cylinder when the piston is at the TDC is called *clearance volume* (V_c). The internal diameter of the cylinder is called *bore.*

There are two classes of reciprocating IC engines:

- *spark-ignition* (or SI) engines — where the combustion of the fuel is initiated by a spark. They are also known as *Otto engines* and use gasoline (petrol) for fuel.
- *compression-ignition* (or CI) engines — where the combustion is initiated spontaneously by virtue of the rise in temperature during the compression process. They are also known as *Diesel engines* and use Diesel fuel.

Consequently, the most common cycles that model the operation of IC engines are: the **Otto cycle**, which models gasoline engines and the **Diesel cycle**, which models diesel engines.

The SI and CI engines are either *two-stroke* or *four-stroke* engines. In the case of the two-stroke engine, a complete cycle is performed during two strokes of the piston inside the cylinder. This means that for every single rotation of the crankshaft an amount of fuel is burnt. In the case of four-stroke engines, a complete cycle is performed during four strokes of the piston inside the cylinder. This means that each time a quantity of fuel is burnt, there are two rotations of the crankshaft.

To simplify the study of the IC engines, a number of assumptions are considered, commonly known as **air-standard assumptions**:

(a) The working fluid is air, in a fixed amount, considered ideal gas; there is no matter exchange with the surroundings, the same working fluid remains in the system during the entire cycle.

(b) All processes are internally reversible.
(c) The combustion process is replaced by a heat-addition process from an external source.
(d) The exhaust process is replaced by a heat-rejection process that restores the fluid to its initial state.

In addition to that, in a **cold air-standard analysis**, the specific heats are assumed constant at their ambient temperature (25 °C) values [7, p. 225]. Since during the operation of IC engines the air–fuel ratio is around 15:1 [kg air/kg fuel], no huge error is made if we consider that the fluid evolving in the cycle has the properties of air (fuel vapor represent only about 6 %).

A full analysis of the performance of IC engines is complex and involves considerations of the combustion process itself. An air-standard analysis simplifies the study of IC engines considerably, but the thermodynamic characteristics of the ideal cycle will differ substantially from those of actual engines. However, useful deductions about the performance of IC engines can be made from this simplified treatment as it can correctly indicate the relative effects of principal variables of the cycle and the upper limit of performance. In this chapter, we consider that the two cycles studied — Otto and Diesel — adhere to cold air-standard cycle idealizations. When the cycles are analyzed on a cold air-standard basis, some important and relatively simple expressions can be obtained for the thermal efficiency.

9.2.1. *The Otto cycle*

The real cycle of an IC engine can be visualized using a device called *indicator*. It records the variation of pressure against volume inside the cylinder and produces a pressure vs. volume plot called *indicator diagram* or *p–V diagram*. Lines or curves on the indicator diagram represent *processes*. The areas under curves are equal to the *work* associated with the process. A typical indicator diagram of a four-stroke Otto engine looks approximately as depicted in Fig. 9.2. In the process 1–2 as the piston moves from the TDC to the BDC, a vacuum is created inside the cylinder and, through

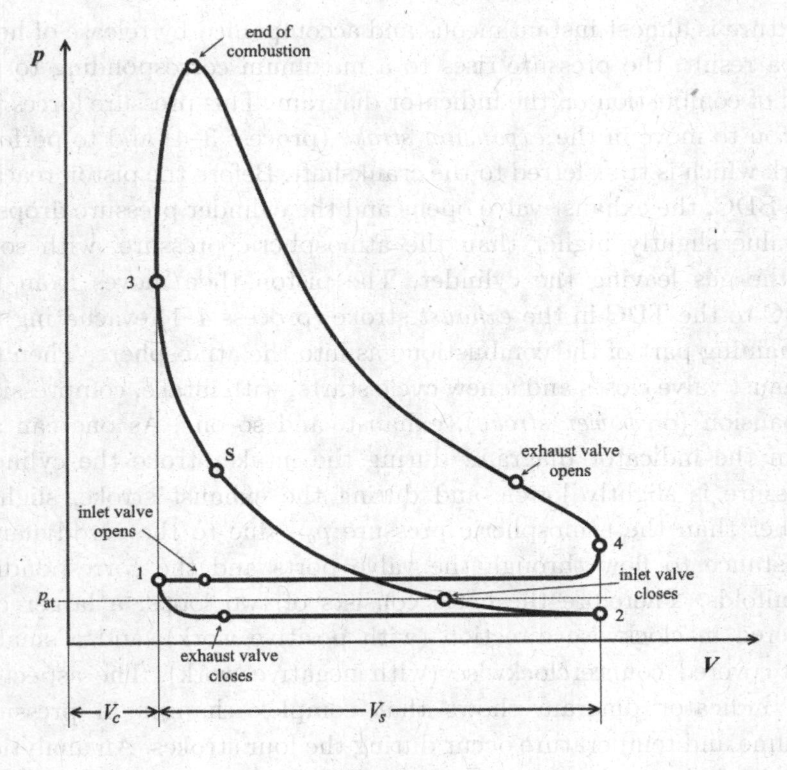

Fig. 9.2

the open inlet valve, the air–fuel mixture prepared in a special device, the carburetor, is absorbed into the cylinder.[b] This is the *intake stroke*. After reaching the BDC, the piston begins to move in the opposite direction and, the process of filling the cylinder with fuel mixture being complete, the inlet valve closes. During this *compression stroke* the pressure of the fuel mixture rises (process 2–3). After the pressure reaches a certain magnitude, corresponding to point S on the indicator diagram, the fuel mixture is ignited by an electric spark produced by a *spark plug*. The combustion of the fuel

[b]Now virtually all modern Otto engines use *indirect* (or *port*) *fuel injection* (fuel is sprayed into the intake manifold behind each intake valve) or cylinder-*direct fuel injection*.

mixture is almost instantaneous and accompanied by release of heat. As a result, the pressure rises to a maximum corresponding to the end of combustion on the indicator diagram. This pressure forces the piston to move in the *expansion stroke* (process 3–4) and to perform work which is transferred to the crankshaft. Before the piston reaches the BDC, the exhaust valve opens and the cylinder pressure drops to a value slightly higher than the atmospheric pressure, with some of the gas leaving the cylinder. The piston then moves from the BDC to the TDC in the *exhaust* stroke (process 4–1) evacuating the remaining part of the combustion gas into the atmosphere. Then the exhaust valve closes and a new cycle starts, with intake, compression, expansion (or *power stroke*), exhaust, and so on.[c] As one can see from the indicator diagram, during the intake stroke the cylinder pressure is slightly lower, and during the exhaust stroke, slightly higher than the atmospheric pressure p_{at}, due to the aerodynamic resistance to flow through the valve ports and the corresponding manifolds. Therefore the cycle consists of two loops: a larger one covered in clockwise direction (with positive work), and a smaller one covered counterclockwise (with negative work). The aspect of the indicator diagram shows that complex changes in pressure, volume and temperature occur during the four strokes. An analytical study of the cycle requires some simplifications, for instance, by considering that the pressures during the intake and exhaust strokes are practically the same. This result in processes 1–2 and 4–1 overlapping and canceling each other out from the standpoint of work exchange; therefore they can be eliminated altogether. Other simplifications are as follows: the compression of gas–fuel mixture in the cylinder is considered adiabatic; the combustion takes place so rapidly at the end of the compression stroke that the volume is essentially constant (isochoric process); the expansion of gases in the cylinder after the ignition of fuel mixture is considered adiabatic (this is the part of the cycle that does positive work); the exhaust of the

[c]The four-stroke IC engine was conceptualized by the French civil engineer, *Alphonse Beau de Rochas* in 1862, and designed and built independently, by the German engineer *Nikolaus August Otto* in 1876.

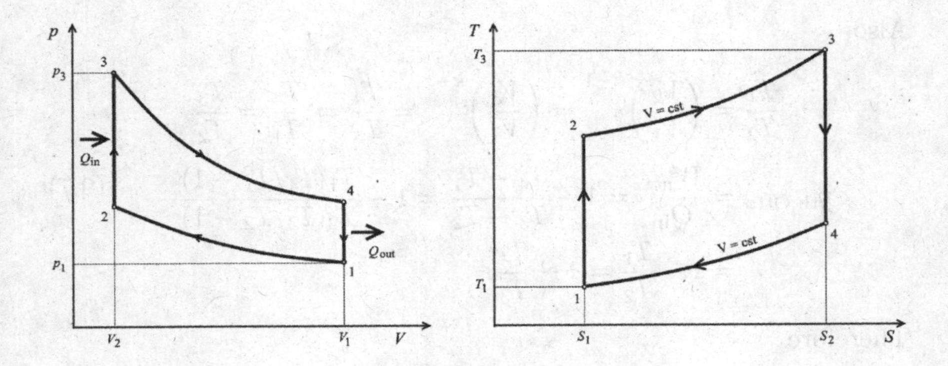

Fig. 9.3

spent gases and the intake of a new fuel mixture into the cylinder are modeled by an isochoric cooling of combustion gases. This results in the ideal cycle represented in Fig. 9.3, known as the *ideal Otto cycle*; it is the same for two- and four-stroke engines. The thermal efficiency of the Otto cycle is

$$\eta_{\text{th,Otto}} = \frac{W_{\text{net}}}{Q_{\text{in}}} \tag{9.1}$$

$$W_{\text{net}} = Q_{\text{in}} - |Q_{\text{out}}| \tag{9.2}$$

$$Q_{\text{in}} = mc_v(T_3 - T_2) \tag{9.3}$$

$$|Q_{\text{out}}| = mc_v(T_4 - T_1) \tag{9.4}$$

Introducing (9.3) and (9.4) in (9.2):

$$\begin{aligned} W_{\text{net}} &= mc_v(T_3 - T_2) - mc_v(T_4 - T_1) \\ &= mc_v[(T_3 - T_2) - (T_4 - T_1)] \end{aligned} \tag{9.5}$$

and

$$\eta_{\text{th,Otto}} = \frac{W_{\text{net}}}{Q_{\text{in}}} = \frac{mc_v[(T_3 - T_2) - (T_4 - T_1)]}{mc_v(T_3 - T_2)} = 1 - \frac{T_4 - T_1}{T_3 - T_2} \tag{9.6}$$

Also,

$$\frac{T_1}{T_2} = \left(\frac{V_2}{V_1}\right)^{k-1} = \left(\frac{V_3}{V_4}\right)^{k-1} = \frac{T_4}{T_3} \Rightarrow \frac{T_4}{T_1} = \frac{T_3}{T_2}$$

$$\begin{aligned}\eta_{\text{th,Otto}} &= \frac{W_{\text{net}}}{Q_{\text{in}}} = 1 - \frac{T_4 - T_1}{T_3 - T_2} = 1 - \frac{T_1(T_4/T_1 - 1)}{T_2(T_3/T_2 - 1)} \\ &= 1 - \frac{T_1}{T_2} = 1 - \frac{T_4}{T_3}\end{aligned} \tag{9.7}$$

Therefore

$$\eta_{\text{th,Otto}} = 1 - \frac{T_4}{T_3} = 1 - \frac{T_1}{T_2} = 1 - \left(\frac{V_2}{V_1}\right)^{k-1} \tag{9.8}$$

Introducing the **compression ratio**, r, defined as

$$r = \frac{V_1}{V_2} \tag{9.9}$$

$$\eta_{\text{th,Otto}} = 1 - \left(\frac{1}{r}\right)^{k-1} \tag{9.10}$$

Equation (9.10) shows that, for a given specific heat ratio k, the thermal efficiency of the Otto cycle increases with the compression ratio r. This conclusion holds true for actual SI IC engines.

In practice, however, it proves impossible to operate SI engines at very high compression ratios because, as compression ratio increases, the temperature of the mixture increases too, reaching the auto-ignition level, when the fuel oxidation process occurs spontaneously in parts or all of the combustion chamber before the release of the spark. This abnormal combustion (accompanied by a specific noise) is called *knocking*, or *detonation*, and has a detrimental effect on the performance and the integrity of the engine. A practical measure of a fuel's resistance to knock is the *octane number*. High octane number fuels are more resistant to knock. Typically the compression ratio for an Otto engine is between 7 and 10. For some special high-performance engines, $r = 12$. Consequently, the thermal efficiency $\eta_{\text{th,Otto}} \approx 54$–$62$ %. In practice, the overall efficiency of an Otto engine is around 35 %.

The work output for an Otto cycle can also be expressed as

$$W = \frac{p_3V_3 - p_4V_4}{k-1} - \frac{p_2V_2 - p_1V_1}{k-1} \tag{9.11}$$

When discussing air-standard cycles for IC engines in general, a useful guide to the relative size of the engine is provided by the concept of **mean effective pressure** (p_m). This conventional pressure is defined as the height of a rectangle on the p–V diagram having the same length ($V_1 - V_2$) and area (W) as the cycle (Fig. 9.4). Thus,

$$p_m = \frac{W_{\text{net}}}{\text{displacement}} = \frac{W_{\text{net}}}{V_{\max} - V_{\min}} = \frac{W_{\text{net}}}{V_1 - V_2} \tag{9.12}$$

This general relation can be written explicitly in many ways using various combinations of state parameters. For example:

$$p_m = \frac{1}{k-1}\left[\frac{V_2(p_3 - p_2) - V_1(p_4 - p_1)}{V_1 - V_2}\right] \tag{9.13}$$

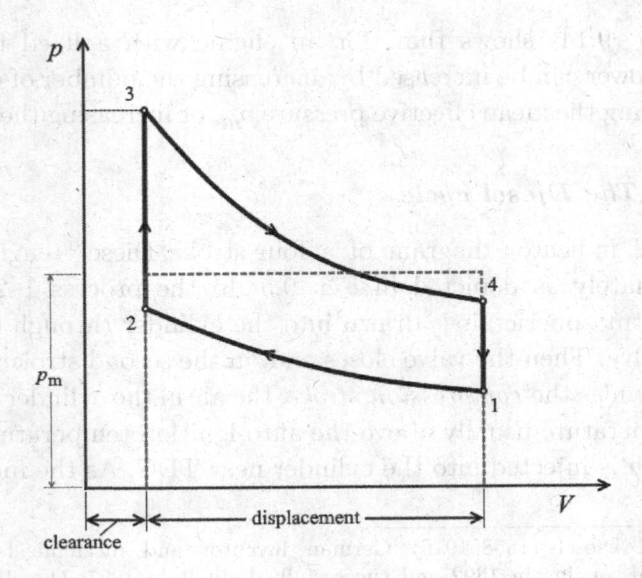

Fig. 9.4

For the cold air-standard cycle of a one-cylinder engine, the power P generated is

$$P = p_m LAN \, [W] \tag{9.14}$$

where the factors are: p_m — mean effective pressure [Pa]; L — piston stroke length [m]; A — piston area [m^2]; N — the number of cycles per unit time [1/s]. Keep in mind that in Eq. (9.14) N is the number of cycles performed by the engine per unit time; do not confuse it with the rotational speed of the engine, n, expressed in [rpm] or [1/min]. If we denote the number of strokes associated to a cycle by τ, the number of revolutions per cycle will be $\tau/2$. Thus

- for a two-stroke engine ($\tau = 2$): $N = \dfrac{n}{\tau/2} = \dfrac{n}{2/2} = n$ [cycles/min]
- for a four-stroke engine ($\tau = 4$): $N = \dfrac{n}{\tau/2} = \dfrac{n}{4/2} = \dfrac{n}{2}$ [cycles/min]

For an engine with i cylinders, the equation of power P becomes:

$$P = i \frac{n}{60\frac{\tau}{2}} W = i p_m LA \frac{n}{60\frac{\tau}{2}} \quad [W] \tag{9.15}$$

Equation (9.14) shows that, for an engine with a fixed bore and stroke, power can be increased by: increasing the number of cylinders i, increasing the mean effective pressure p_m, or increasing the speed n.

9.2.2. *The Diesel cycle*

A typical indicator diagram of a four-stroke Diesel[d] engine looks approximately as depicted in Fig. 9.5. In the process 1–2 (*intake stroke*) atmospheric air is drawn into the cylinder through the open intake valve. Then the valve closes and, in the second stroke (process 2–3) known as the *compression stroke*, the air in the cylinder heats up to a temperature usually above the auto-ignition temperature of the fuel which is injected into the cylinder near TDC. As the fuel burns,

[d]Rudolph Diesel (1858–1916) German inventor and mechanical engineer, patented his engine in 1892 and successfully built it in 1897. Diesel originally designed his engine to use coal dust as fuel, but experimented with other fuels including peanut oil.

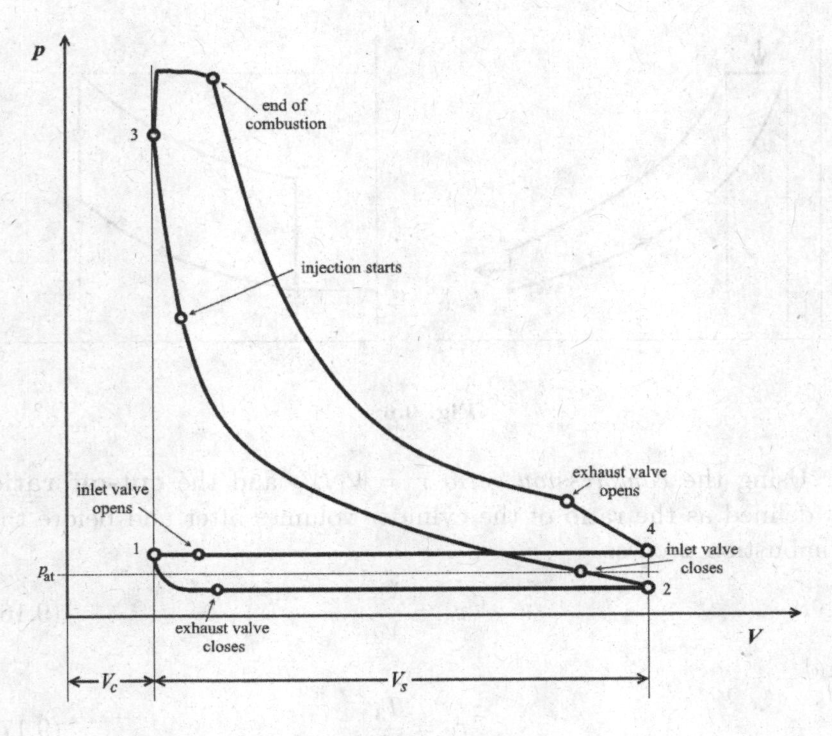

Fig. 9.5

heat is generated raising the pressure inside the combustion chamber near TDC. This pressure is applied to the piston thus pushing it back toward the BDC as the fuel continues to burn for a period of time practically at constant pressure. This stroke (process 3–4) is called *expansion stroke* and is also known as the *power stroke* for obvious reasons. The last of the four strokes is the *exhaust stroke* (process 4–1) where combustion byproducts are evacuated into the atmosphere through the exhaust valve. The idealization of the cycle leads to the diagrams in Fig. 9.6. As one can see, the only difference between the ideal Otto cycle and the *ideal Diesel cycle* is the combustion process, which occurs theoretically at constant pressure for the Diesel cycle, and at constant volume for the Otto cycle. The Diesel cycle is, therefore, referred to as the *constant-pressure combustion cycle*, while the Otto cycle is referred to as the *constant-volume combustion cycle.*

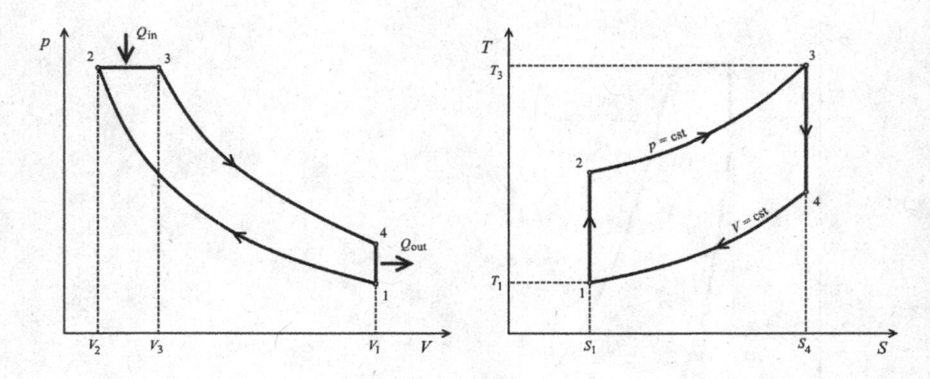

Fig. 9.6

Using the *compression ratio* $r = V_1/V_2$ and the **cut-off ratio** r_c defined as the ratio of the cylinder volumes after and before the combustion process,

$$r_c = \frac{V_3}{V_2} \tag{9.16}$$

and

$$r_c = \frac{T_3}{T_2} \tag{9.17}$$

the thermal efficiency of the Diesel cycle becomes

$$\eta_{\text{th,Diesel}} = \frac{Q_{\text{in}} - |Q_{\text{out}}|}{Q_{\text{in}}} = \frac{mc_p(T_3 - T_2) - mc_v(T_4 - T_1)}{mc_p(T_3 - T_2)}$$

$$\eta_{\text{th,Diesel}} = 1 - \frac{c_v(T_4 - T_1)}{c_p(T_3 - T_2)} = 1 - \frac{1}{k}\left(\frac{T_4 - T_1}{T_3 - T_2}\right) \tag{9.18}$$

Also

$$T_3 = T_2 r_c = T_1 r^{(k-1)} r_c \tag{9.19}$$

$$T_2 = T_1 \left(\frac{V_1}{V_2}\right)^{k-1} = T_1 r^{k-1} \tag{9.20}$$

$$T_4 = T_3 \left(\frac{V_3}{V_4}\right)^{k-1} = T_3 \left(\frac{V_3}{V_2} \cdot \frac{V_2}{V_4}\right)^{k-1} = T_3 \left(\frac{r_c}{r}\right)^{k-1}$$

$$T_4 = T_1 r^{(k-1)} r_c \left(\frac{r_c}{r}\right)^{k-1} = T_1 r_c^k \tag{9.21}$$

Therefore

$$\eta_{\text{th,Diesel}} = 1 - \frac{1}{k}\left[\frac{T_1(r_c^k - 1)}{T_1(r^{k-1}r_c - r^{k-1})}\right] = 1 - \frac{1}{k}\left[\frac{r_c^k - 1}{r^{k-1}r_c - r^{k-1}}\right]$$

$$\eta_{\text{th,Diesel}} = 1 - \frac{1}{r^{k-1}}\left[\frac{r_c^k - 1}{k(r_c - 1)}\right] \tag{9.22}$$

It may be noted that the difference between the efficiencies of the Diesel and Otto cycle is in the bracketed factor. Because $r_c > 1$ and $k = 1.4$ (air), then the bracket factor is greater than 1. So the efficiency of the Diesel cycle is smaller than that of the Otto cycle for a given compression ratio.

However, this advantage of the Otto cycle is of no practical value, because the compression ratio of this cycle is limited to 7–10, whereas for the Diesel cycle is 16–25. Due to its higher compression ratio, the actual Diesel engine is more efficient than a gasoline engine.

The net work output for a Diesel cycle is

$$W = p_2(V_3 - V_2) + \frac{p_3V_3 - p_4V_4}{k-1} - \frac{p_2V_2 - p_1V_1}{k-1}$$

$$W = p_2V_2(r_c - 1) + \frac{p_3r_cV_2 - p_4rV_2}{k-1} - \frac{p_2V_2 - p_1rV_2}{k-1}$$

$$W = V_2\left[\frac{p_2(r_c-1)(k-1) + p_3r_c - p_4r - (p_2 - p_1r)}{k-1}\right]$$

$$W = V_2\left[\frac{p_2(r_c-1)(k-1) + p_3\left(r_c - \frac{p_4}{p_3}r\right) - p_2\left(1 - \frac{p_1}{p_2}r\right)}{k-1}\right]$$

$$W = p_2V_2\left[\frac{(r_c-1)(k-1) + (r_c - r_c^k r^{1-k}) - (1 - r^{1-k})}{k-1}\right]$$

$$W = p_1V_1r^{k-1}\frac{k(r_c-1) - r^{1-k}(r_c^k - 1)}{k-1} \tag{9.23}$$

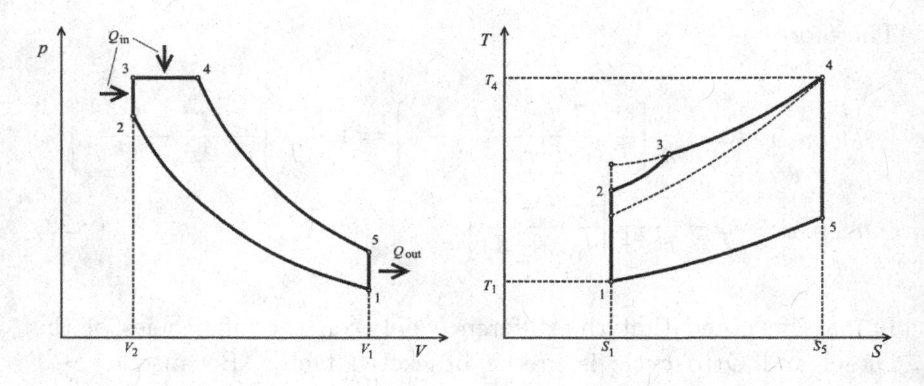

Fig. 9.7

The mean effective pressure is

$$p_m = p_1 V_1 \frac{r^{k-1} k(r_c - 1) - (r_c^k - 1)}{(k-1) V_1 \frac{r-1}{r}}$$

$$p_m = p_1 \frac{k r^k (r_c - 1) - r(r_c^k - 1)}{(k-1)(r-1)} \tag{9.24}$$

9.2.3. *The dual cycle*

The *dual cycle* is sometimes used to approximate actual IC engine cycles since, for the Otto cycle, the time necessary for the heat input is not zero (combustion extends into the expansion stroke) and for the Diesel cycle, due to the nature of the combustion process, the heat input does not occur only at constant pressure. Thus, in a dual cycle heat Q_{in}, is first added along the isochore 2–3, then along the isobar 3–4 (Fig. 9.7). The dual cycle, also called a *mixed cycle*, is a compromise between Otto and Diesel cycles. An example of thermodynamic study of a dual cycle can be found in [37]. In Fig. 9.7, the dotted lines represent the isobar corresponding to point 3 and the isochore corresponding to point 4, respectively.

9.3. Other types of IC engines

As mentioned in Sec. 9.2.1, in actual four-stroke engines during the intake and exhaust strokes the engine performs negative work. An improvement is offered by the solution of **two-stroke engines**. In this case the entire working cycle is produced in two strokes. The intake and exhaust strokes are eliminated, since the exchange of fluids (a fresh load replacing the combustion products) takes place at the end of the expansion stroke through special ports opened by the moving piston; the crankcase participates also to the cycle. Two-stroke engines can be built as CI or SI engines; they use the same ideal cycles as their corresponding four-stroke engines. Details about the operation of two-stroke engines can be found in [1, p. 397; 11, p. 505]. Although traditional two-stroke engines produce more pollution because of the combustion of the oil in the gas and because each time a new mix of air/fuel is loaded into the combustion chamber, part of it leaks out through the exhaust port, new two-stroke engines are associated with low level of pollutant emissions [38]. Most of the current applications are in marine practice, as CI engines.

The **Atkinson cycle engine** is a type of internal combustion engine invented by James Atkinson in 1882. A special arrangement of levers allows the Atkinson engine to perform one cycle (four strokes) in only one revolution of the main crankshaft, with the compression and expansion strokes asymmetrical. The design eliminates the need for a separate cam shaft. The Atkinson cycle is designed to operate with high efficiency but produces lower power output. It is used in some modern hybrid electric applications [39] (e.g., 2014 Honda Accord Hybrid, Toyota Aygo).

The **Wankel engine**[e] (*rotary engine*) [1, p. 398] does not have reciprocating pistons. It operates based on the Otto cycle with the same separation of phases as a four-stroke engine. All phases take place in the space between the sides of the rotor (which is triangular with bow-shaped flanks) and the epitrochoid-shaped housing surrounding the rotor (Fig. 9.8 [40]). Wankel engines are considerably simpler and contain fewer moving parts: there are no valves, no camshafts; the rotor is geared directly to the output shaft, so no need for connecting rods, conventional crankshaft, crankshaft

[e]Named after Felix Heinrich Wankel (Germany, 1902–1988), its inventor, who designed the engine in 1954 and tested it in 1957.

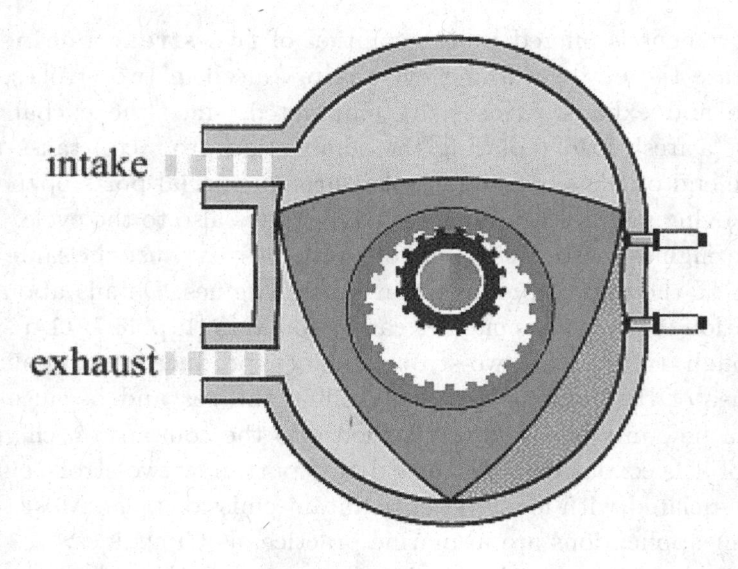

Fig. 9.8

balance weights, etc. [41]. Therefore a Wankel engine is much lighter (typically half that of a conventional engine with equivalent power) and completely eliminates the reciprocating mass of a piston engine with its internal strain and inherent vibration. It is able to produce more power by running at higher speed resulting in a greater power-to-weight ratio than piston engines. The configuration of a rotary engine creates difficulties with seals, hydrocarbon emissions and fuel efficiency. These problems seem to have been solved by the car manufacturer Mazda in their current RX-8 model and the earlier RX-7.

9.4. Stirling and Ericsson cycles

The Otto, Diesel and mixed cycles generate a high amount of net work per cycle but are all less efficient than a Carnot cycle operating the same maximum and minimum temperatures. The Carnot cycle, though, has the deficiency of having a small mean effective pressure, which means, when used for a reciprocating engine, it produces little

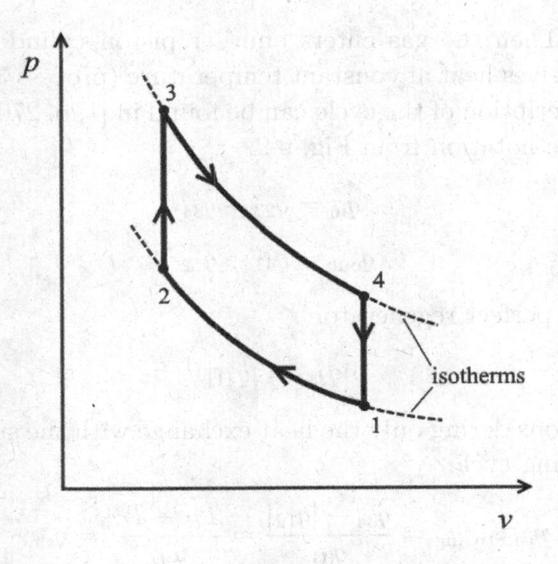

Fig. 9.9

work for a given displacement. There are other cycles that do not have this downside and still have a Carnot efficiency: the Stirling cycle and the Ericsson cycle.

The **Stirling cycle** has the Carnot efficiency and a much higher mean effective pressure. It consists of two isochoric processes and two isothermals (see Fig. 9.9). The processes are as follows: 1–2, isothermal compression; 2–3, isochoric heat addition; 3–4, isothermal expansion; 4–1, isochoric regeneration. The heat supplied during process 2–3 is quantitatively equal to the heat rejected during process 4–1. This is accomplished reversibly in a *regenerator* [1, p. 270], which consists of a matrix of wire gauze or a ceramic mesh, or any kind of porous material with high heat capacity.[f] An amount of hot gas enters the matrix in state 4 and leaves in state 1 releasing some heat to the matrix. Then the gas releases heat in a piston–cylinder apparatus at constant temperature (process 1–2). In state 2, the gas passes in reversed direction through the matrix, cooling it, and leaving

[f]Heat capacity C is equal to the product between mass and specific heat: $C = mc$.

in state 3. Then the gas enters another piston–cylinder apparatus where it receives heat at constant temperature (process 3–4). A more detailed description of the cycle can be found in [1, p. 270; 11, p. 514].

Using the notation from Fig. 9.9,

$$q_{\text{in}} = q_{23} + q_{34}$$

$$q_{\text{out}} = q_{41} + q_{12}$$

But, with a perfect regenerator,

$$|q_{23}| = |q_{41}|$$

Therefore, considering only the heat exchange with the surroundings, for the Stirling cycle

$$\eta_{\text{th,Stirling}} = \frac{q_{34} - |q_{12}|}{q_{34}} = \frac{T_H - T_C}{T_H} = \eta_{\text{th},C} \tag{9.25}$$

An engine, operating based on the Stirling[g] cycle, is not in fact an IC engine, and is sometimes referred to as *external combustion engine*. The advantage of such an engine is that any fuel can be used to produce heat. Although efficient and developing a mean effective pressure comparable to that of the Otto cycle, the Stirling engine did not gain popularity because it is heavy and expensive. Modern developments use high-pressure helium or hydrogen as working gas.

The **Ericsson cycle** [1, p. 272; 11, p. 515] consists of two isobars and two isotherms (Fig. 9.10). The processes are as follows: 1–2, isothermal compression; 2–3, isobaric heat-addition; 3–4, isothermal expansion; 4–1, isobaric heat-rejection. In this case, the heat rejected during one isobaric process is restituted to the working fluid during the other isobaric process, via a regenerator. An Ericsson cycle for gas-turbine power plants is represented in Fig. 9.11. The advantage of the Ericsson cycle[h] over the Carnot and Stirling cycles is its smaller pressure ratio for a given ratio of maximum to minimum specific

[g]Reverend Dr. Robert Stirling (1790–1878), a Scottish clergyman and inventor of the Stirling engine. He patented his hot air engine in 1816.

[h]Named after its inventor, John Ericsson (1803–1889), a Swedish-American inventor and mechanical engineer.

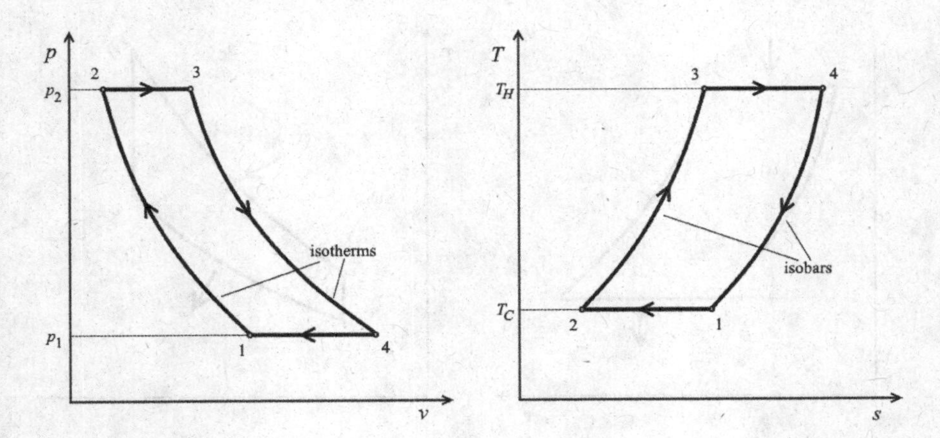

Fig. 9.10

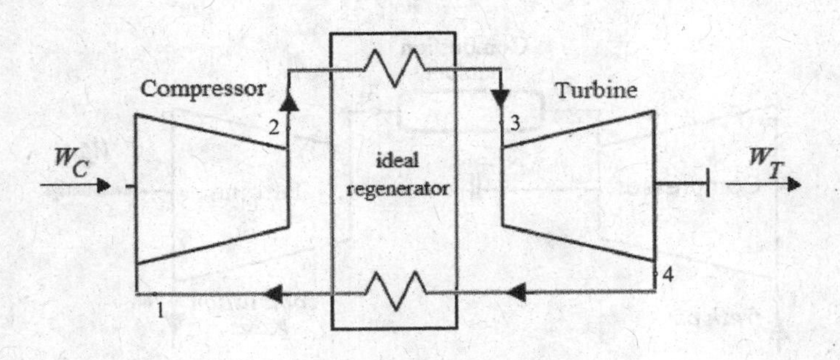

Fig. 9.11

volume with higher mean effective pressure [42, p. 80]. The thermal efficiency is, as in the case of the Stirling cycle,

$$\eta_{\text{th,Ericsson}} = \frac{T_H - T_C}{T_H} = \eta_{\text{th},C} \tag{9.26}$$

9.5. The Brayton cycle

The **Brayton cycle** is the ideal cycle for gas-turbine engines. It consists of four processes, as shown in Fig. 9.12:

1–2: adiabatic compression (in a compressor);
2–3: isobaric heat-addition (in a combustion chamber);

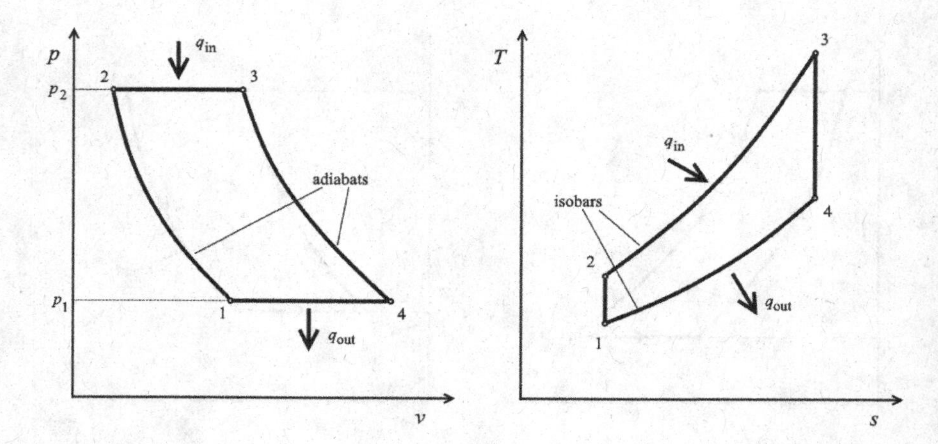

Fig. 9.12

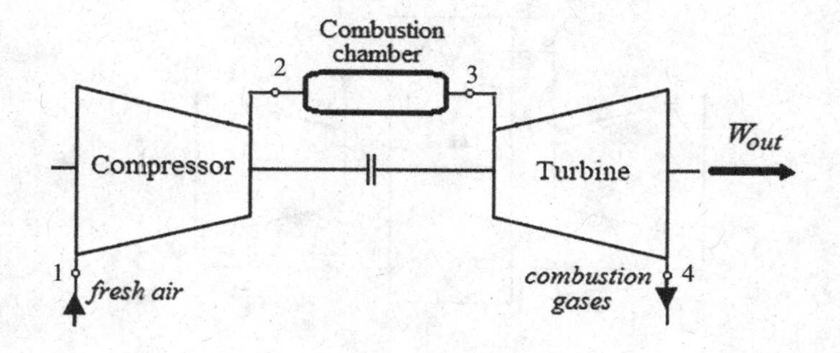

Fig. 9.13

3–4: adiabatic expansion (in a turbine);
4–1: isobaric heat-rejection.

In the ideal open-cycle this last process replaces the actual evacuation of combustion gases in the ambient and the intake of fresh air (Fig. 9.13). Applying the cold air-standard assumptions, for each component of the Brayton cycle the steady-flow energy equation for one unit mass can be written as

$$\Sigma q - \Sigma w = \Delta h$$
$$(q_{\text{in}} + q_{\text{out}}) - (w_{\text{in}} + w_{\text{out}}) = h_{\text{final}} - h_{\text{initial}}$$

Therefore, the heat exchanges with the surroundings are

$$q_{\text{in}} = h_3 - h_2 = c_p(T_3 - T_2)$$
$$q_{\text{out}} = h_1 - h_4 = c_p(T_1 - T_4)$$

The thermal efficiency is

$$\begin{aligned}
\eta_{\text{th,Brayton}} &= \frac{q_{\text{in}} - |q_{\text{out}}|}{q_{\text{in}}} = \frac{(h_3 - h_2) - |h_1 - h_4|}{h_3 - h_2} \\
&= \frac{(T_3 - T_2) - (T_4 - T_1)}{T_3 - T_2}
\end{aligned} \tag{9.27}$$

$$\eta_{\text{th,Brayton}} = 1 - \frac{T_4 - T_1}{T_3 - T_2} = 1 - \frac{T_1\left(\frac{T_4}{T_1} - 1\right)}{T_2\left(\frac{T_3}{T_2} - 1\right)} \tag{9.28}$$

Processes 1–2 and 3–4 being isentropic, and $p_1 = p_4$ and $p_2 = p_3$,

$$\frac{T_2}{T_1} = \left(\frac{p_2}{p_1}\right)^{(k-1)/k} = \left(\frac{p_3}{p_4}\right)^{(k-1);k} = \frac{T_3}{T_4}$$

Substituting in (9.28), and defining the **pressure ratio** as $r_p = \frac{p_2}{p_1}$,

$$\eta_{\text{th,Brayton}} = 1 - \frac{1}{r_p^{(k-1)/k}} \tag{9.29}$$

Equation (9.29) shows that, under the cold air-standard assumptions and for a given specific heat ratio k, the thermal efficiency of the ideal Brayton cycle increases with the pressure ratio r_p. This conclusion holds also for actual gas turbines. An increase in r_p, however, increases the maximum temperature of the cycle T_3. With higher r_p values, the fuel flow rate has to be reduced to keep T_3 within acceptable limits. Therefore, currently, the economic optimum is around $r_p \approx 15$.

The back work ratio r_{bw} for the cycle is

$$r_{\text{bw}} = \frac{|w_C|}{w_T} = \frac{h_2 - h_1}{h_3 - h_4} \tag{9.30}$$

It is to be noticed that for gas-turbine plants r_{bw} is typically 40–80 % [7, p. 237], much higher than for steam-turbine plants.

The Brayton cycle[i] is also known as the *Joule cycle.*

9.6. Jet-propulsion cycle

In the 1930s and 1940s gas-turbine engines began to be used to power aircraft, first under military sponsorship. Gradually they became the norm because of their advantages over piston IC engines: reduced weight, compactness, and high power-to-weight ratio. In principle, aircraft are propelled by accelerating a fluid in the opposite direction of motion [11, p. 531]. The propulsive work produced by aircraft power plant is used either to slightly accelerate a large mass of fluid — as in the case of *propeller-driven engines* — or to significantly accelerate a small mass of fluid — as in the case of *jet* or *turbojet engines.* A combination of both characterizes the operation of *turboprop engines.* In this section we will discuss only the jet engines.

Aircraft gas turbines operate on an open cycle called a **jet-propulsion cycle**. This ideal cycle (Fig. 9.14) is similar to a simple Brayton cycle with the difference that the gases are not expanded in the turbine to atmospheric pressure, but to a higher pressure, just enough to cover the energy consumed by the compressor (and some auxiliary equipment). The gases exiting the turbine at relatively high pressure are then accelerated in a nozzle to produce the **thrust**[j] that propels the aircraft. Therefore, the net work output of a jet-propulsion cycle is zero. Thus, the traditional concept of thermal efficiency is meaningless for turbojet engines.

Turbine engines come in a variety of forms, but all of them have the same core components: compressor, burner section, and turbine which drives the compressor. The role of the diffuser (Fig. 9.15) is to transform the kinetic energy of the entering air into potential

---

[i]Named after George Brayton (1830–1892) American mechanical engineer, who first proposed the concept in 1873.

[j]Thrust is the unbalanced force caused by the difference in momentum of the jet relative to the aircraft and the air entering the engine.

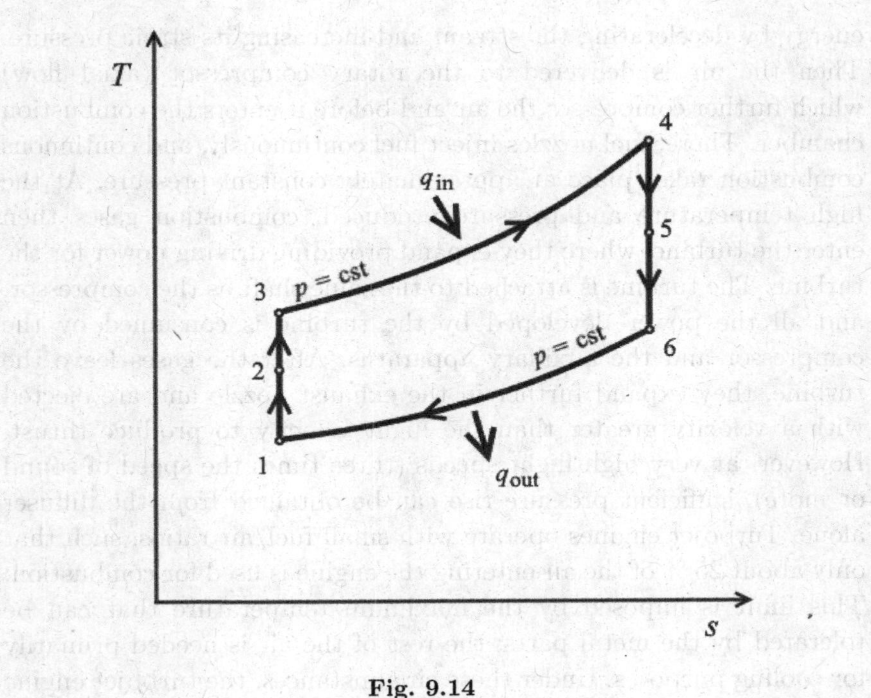

Fig. 9.14

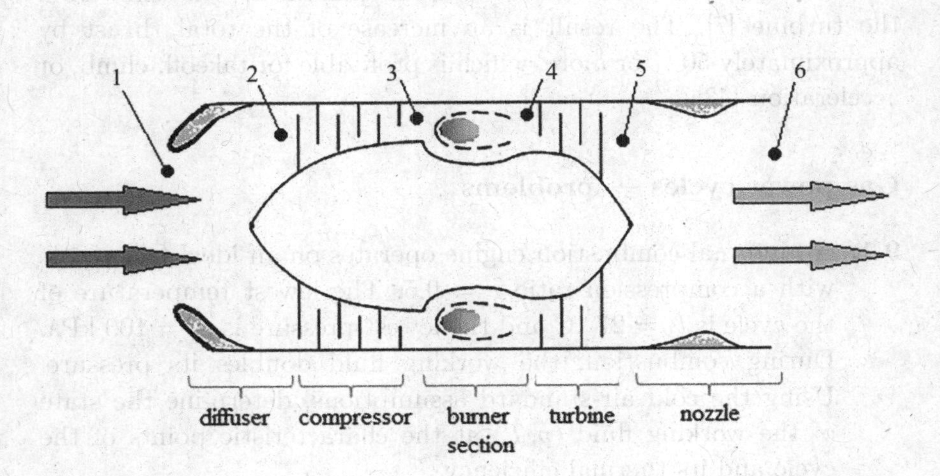

Fig. 9.15

energy by decelerating the stream and increasing its static pressure. Then the air is delivered to the rotary compressor (axial flow) which further compresses the air and before it enters the combustion chamber. There, fuel nozzles inject fuel continuously, and continuous combustion takes place at approximately constant pressure. At the high temperature and pressure produced, combustion gases then enter the turbine, where they expand providing driving power for the turbine. The turbine is attached to the same shaft as the compressor, and all the power developed by the turbine is consumed by the compressor and the auxiliary apparatus. After the gases leave the turbine, they expand further in the exhaust nozzle and are ejected with a velocity greater than the flight velocity to produce thrust. However, at very high flight speeds (three times the speed of sound or more), sufficient pressure rise can be obtained from the diffuser alone. Turbojet engines operate with small fuel/air ratios, such that only about 25 % of the air entering the engine is used for combustion. This limit is imposed by the maximum temperature that can be tolerated by the metal parts; the rest of the air is needed primarily for cooling purposes. Under these circumstances, the turbojet engine can be equipped with an afterburner placed at the exit from the turbine [7]. The result is an increase of the total thrust by approximately 50 % or more, which is profitable for takeoff, climb, or acceleration [43].

Gas power cycles — problems

9.1. An internal combustion engine operates on an ideal Otto cycle with a compression ratio $r = 9.5$. The lowest temperature of the cycle is $t_1 = 27$ °C and the lowest pressure is $p_1 = 100$ kPa. During combustion, the working fluid doubles its pressure. Using the cold air-standard assumptions, determine the state of the working fluid (p, T) at the characteristic points of the cycle and its thermal efficiency.

9.2. A supercharged internal combustion engine operates on an ideal Diesel cycle with a compression ratio $r = 20$. The lowest temperature of the cycle is $t_1 = 200$ °C and the lowest pressure is $p_1 = 200$ kPa. Determine the thermal efficiency and the mean effective pressure if the maximum temperature of the cycle is 3200 K and the work output is 1200 kJ/kg.

9.3. An internal combustion engine operates on the cold air-standard dual cycle. At the start of the compression stroke, conditions in the cylinder are: pressure $p_1 = 0.85$ bar and temperature $t_1 = 50$ °C. The compression ratio is $r = 8$, the cut-off ratio $r_c = 1.2$ and the pressure ratio $r_p = 2$. Determine: (a) the state of the working fluid (p, v, T) at the characteristic points of the cycle; (b) the heat added during the cycle; (c) the net work of the cycle; (d) the thermal efficiency of the cycle.

9.4. A Stirling cycle operates on air with a compression ratio of 6. If the low pressure is 0.1 MPa, the low temperature is 25 °C and the high temperature is 1100 °C, determine the work output and the heat input, the maximum pressure and the efficiency of the cycle.

9.5. An Ericsson cycle operates on air, at a mass flow rate of 1 kg/s. Air is compressed and heated from $p_1 = 100$ kPa and $t_1 = 100$ °C to the turbine inlet conditions $t_3 = 600$ °C and $p_3 = 1000$ kPa. Determine the power output, the heat input and the thermal efficiency.

9.6. A power plant operates on a cold air-standard Brayton cycle. Air enters the compressor of the gas turbine at $t_1 = 27$ °C and $p_1 = 100$ kPa. For a pressure ratio of 6.5 and a maximum temperature of 800 °C, determine the back work ratio and the thermal efficiency of the cycle.

Chapter 10

Refrigeration Cycles

When a thermodynamic system undergoes a cycle, the work is done *by* the system *on* the surroundings if the state changes happen in a clockwise manner (power cycles). Conversely, the work is done *on* the system if the state changes happen in an anticlockwise manner. In this case heat is received at low temperature and rejected at high temperature while work is done on the fluid. These are called **refrigeration cycles**. The fluid used in these cycles is called **refrigerant**.

10.1. Refrigerators and heat pumps

The thermal machines operating in a refrigeration cycle are called refrigerators or heat pumps. The distinction in terminology is arbitrary, because heat pumps and refrigerators are identical in principle. If the primary purpose of the cycle is the cooling effect (extraction of heat from a cold space), the device is called **refrigerator** and it operates in a *refrigeration cycle.* If the primary purpose is the heating effect (supply of heat to a warm space), the device is called a **heat pump** and it operates in a *heat pump cycle.* For these devices, the reference cycle is the reversed Carnot cycle.

10.2. The reversed Carnot cycle

When reversing a Carnot cycle, heat is introduced at low temperature and rejected at high temperature while work is added to the system

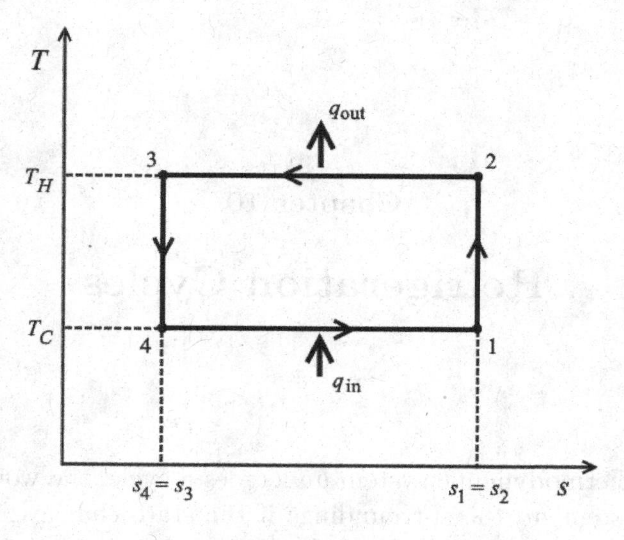

Fig. 10.1

(Fig. 10.1). A reversed Carnot cycle also has superiority of the performance index over real cycles. For the reversed cycle, the performance index is the **coefficient of performance** (COP):

$$\text{COP} = \frac{\text{Desired output}}{\text{Required input (work input)}}$$

The "desired output" will be:

- the heat extracted from the cold space Q_{in} — for refrigerators,
- the heat supplied to the warm space Q_{out} — for heat pumps.

The COP is denoted by β for refrigerators, and by γ for heat pumps. Therefore, for a refrigerator operating on a Carnot cycle

$$\text{COP}_{R.C} = \beta_C = \frac{q_{\text{in}}}{|w_{\text{net.in}}|} = \frac{T_C\,\Delta s}{(T_H - T_C)\,\Delta s}$$

$$= \frac{T_C}{T_H - T_C} = \frac{1}{\frac{T_H}{T_C} - 1} \tag{10.1}$$

and for a heat pump operating on a Carnot cycle

$$\mathrm{COP}_{HP,C} = \gamma_C = \frac{|q_{\text{out}}|}{|w_{\text{net.in}}|} = \frac{T_H\,\Delta s}{(T_H - T_C)\,\Delta s}$$

$$= \frac{T_H}{T_H - T_C} = \frac{1}{1 - \frac{T_C}{T_H}} \tag{10.2}$$

It is to be noticed from Eqs. (10.1) and (10.2) that COP is greater than unity (COP > 1), as opposed to the concept of efficiency ($\eta_{\text{th}} < 1$). Also, for the same temperature limits, T_H and T_C,

$$\mathrm{COP}_{HP,C} = \mathrm{COP}_{R,C} + 1 \tag{10.3}$$

For a refrigeration system, the rate of heat removal from the refrigerated space $\dot{Q}_{\text{in}}$ is referred to as the **refrigeration capacity** [7, p. 210]. This is expressed in W (in SI) or Btu/min (US Customary Units). In the US it is common to measure the refrigeration capacity in *tons of refrigeration.* One ton of refrigeration (US) is the capacity of a refrigeration system to freeze one ton (2000 lb) of liquid water at 0 °C (32 °F) to ice at 0 °C in 24 hours.

The specific latent heat of fusion for ice is 334.5 kJ/kg or 144 Btu/lb. Therefore:

$$\begin{aligned} 1 \text{ ton of refrig.} &= 908 \text{ kg} \times 334.5 \text{ kJ/kg} \\ &\approx 303\,700 \text{ kJ/day (3.52 kW)} \end{aligned}$$

$$\begin{aligned} 1 \text{ ton of refrig.} &= 2000 \text{ lb} \times 144 \text{ Btu/lb} \\ &= 288\,000 \text{ Btu/day (200 Btu/min)} \end{aligned}$$

10.3. The ideal vapor-compression cycle

A reversed Carnot cycle that uses a refrigerant of some sort is represented in Fig. 10.2. The cycle consists of two adiabats and two isotherms. The processes are:

1–2 isentropic (reversible adiabatic) compression;
2–3 isothermal heat rejection (condensation);
3–4 isentropic (reversible adiabatic) expansion;
4–1 isothermal heat absorption (evaporation).

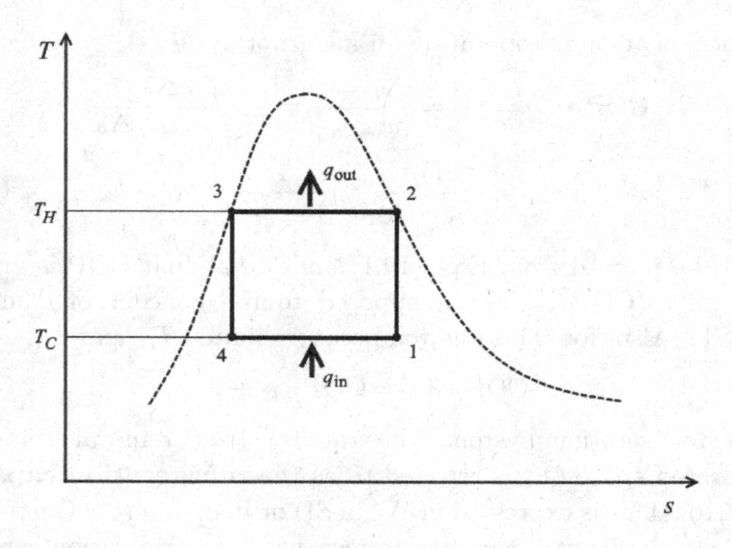

Fig. 10.2

Although this perfectly reversible cycle is associated with the highest COP, it is affected by impracticalities: process 1–2 involves the compression of a mixture of liquid and gas until all the liquid has evaporated (compression of wet mixtures is very difficult to implement mechanically); it is difficult to stop processes exactly at the saturation limit; expansion from 3–4 occurs in a saturated region with very wet refrigerant (low quality), causing erosion of turbine blades. The solution consists of vaporizing the refrigerant completely before it is compressed and replacing the turbine by a throttling valve. This way the impracticalities are eliminated, but the resulting cycle, called *ideal vapor compression cycle* is no longer reversible. This cycle consists of (see Figs. 10.3 and 10.4):

1–2: isentropic compression of the refrigerant in gaseous state (*compressor*);
2–3: condensation at constant pressure (*condenser*);
3–4: expansion through a *throttling valve* (approximately an isenthalpic process, $q = w = \Delta h = 0$);
4–1: evaporation of the refrigerant (*evaporator*).

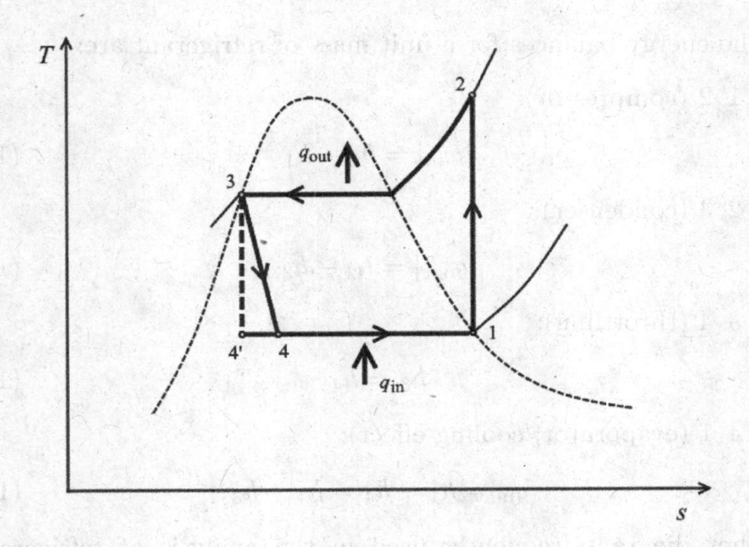

Fig. 10.3

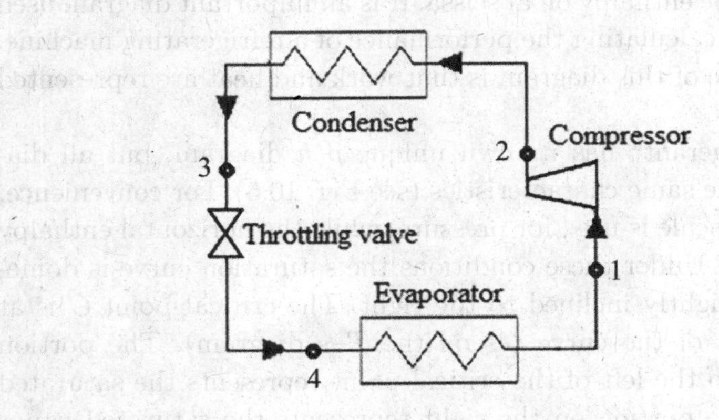

Fig. 10.4

The ideal refrigeration cycle is not internally reversible due to the presence of the throttling process 3–4. For complete reversibility, an isentropic expansion in a turbine 3–4′ should be introduced but that is, as mentioned before, totally impractical. Besides, with a throttling valve it is easy to adjust the pressure of fluid and the lowest temperature of the cycle.

The energy balances for a unit mass of refrigerant are:

path 1–2 (compressor):

$$|w_{\text{in}}| = h_2 - h_1 \tag{10.4}$$

path 2–3 (condenser):

$$|q_{\text{out}}| = h_2 - h_3 \tag{10.5}$$

path 3–4 (throttling):

$$h_3 = h_4 \tag{10.6}$$

path 4–1 (evaporator; cooling effect):

$$q_{\text{in}} = h_1 - h_4 = h_1 - h_3 \tag{10.7}$$

Another diagram frequently used in the analysis of refrigeration cycles is the *p–h* diagram, having the absolute pressure in ordinate and the specific enthalpy on abscissa. It is an important diagram used frequently for calculating the performance of a refrigerating machine. The advantage of this diagram is that work and heat are represented by segments.

Each refrigerant[a] has its own unique *p–h* diagram, but all diagrams have the same characteristics (see Fig. 10.5). For convenience, a logarithmic scale is used for pressure, while the horizontal enthalpy scale is linear. Under these conditions the saturation curve is dome-shaped and slightly inclined to the right. The critical point C is at the maximum of the curve (as on the *T–s* diagram). The portion of the curve to the left of the critical point represents the saturated liquid line, the portion on the right represents the saturated vapor line, and anywhere inside the dome the refrigerant is a mixture of saturated liquid and saturated vapor. Lines of constant pressure are horizontal and lines of constant enthalpy are vertical. Lines of equal

[a] As in the case of the Rankine cycle, in a refrigeration cycle the working fluid undergoes phase changes. Therefore real gas/liquid property tables and charts are required. In all industrial steam power plants the working fluid is water, so only water tables/diagrams are needed. For refrigeration, various refrigerants exist and each one requires its own set of tables and charts.

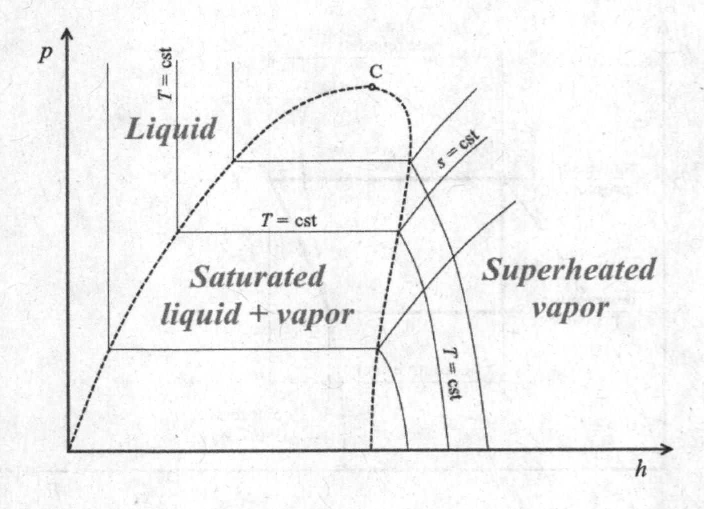

Fig. 10.5

temperature (isotherms) are vertical in the subcooled liquid region. In the saturated liquid–vapor region, isotherms become horizontal (like isobars) illustrating that the saturation temperature is constant at a given pressure. Once the isotherms leave the liquid–vapor dome, they have a downward slope until they become again vertical at some distance from the saturation state. In the superheated vapor and sub-cooled liquid portion of the chart temperatures increase with enthalpy. The lines of constant entropy are represented in the superheated vapor region; they have positive slope.

If the operating temperatures and pressures are known, the refrigeration cycle can be plotted on the p–h diagram. The saturated vapor enters the *compressor* in state 1 in Fig. 10.6. The vapor is compressed, along a constant entropy line to the pressure corresponding to state 2 in the superheated region. At this point, condensation begins, in an isobaric process. As heat continues to be transferred from the refrigerant vapor to the air (or another coolant used in the *condenser*) the refrigerant condenses at constant pressure until it turns into saturated liquid (state 3). Now the refrigerant is in the liquid state. The liquid at a high pressure and temperature then moves to the *throttling valve.* This valve restricts the flow of the fluid and lowers

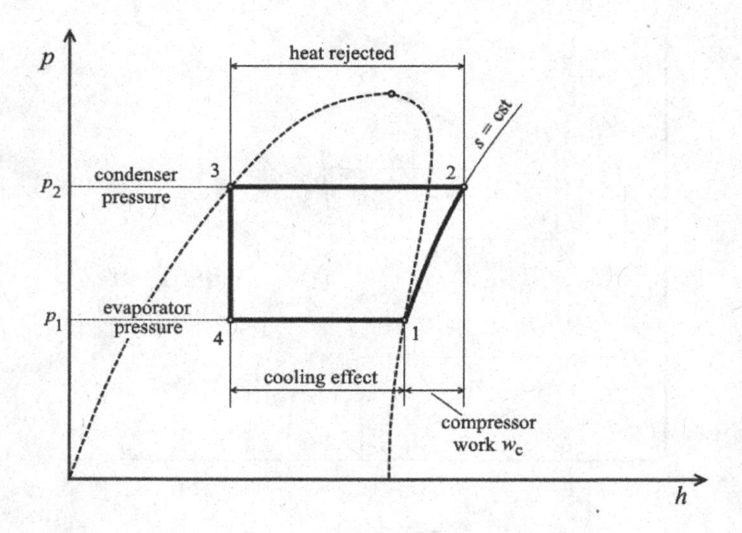

Fig. 10.6

its pressure. The throttling process occurs at constant enthalpy. The low-pressure liquid (state 4) then moves to the *evaporator*, where heat from a higher-temperature medium is absorbed and changes the phase of the refrigerant from a liquid to a saturated gas (state 1). If the device operates as a refrigerator, the evaporator is placed inside the refrigerated space and the condenser outside. If it operates as a heat pump, the condenser is placed inside the heated room and the evaporator outside.

The function of the heat pump may be compared to that of a water pump positioned between two water reservoirs that are located at different altitudes. Although water naturally flows from the higher to the lower reservoir, it is possible to return water to the higher reservoir by using a pump, which draws water from the lower one.

It can be seen from Fig. 10.6 that the amounts of heat and work associated with the cycle is reduced to a simple measurement of segment length in the direction of enthalpy axis. Thermodynamic properties of some refrigerants can be found in [7] as well as in Appendix E (Table A.6, Fig. A.7).

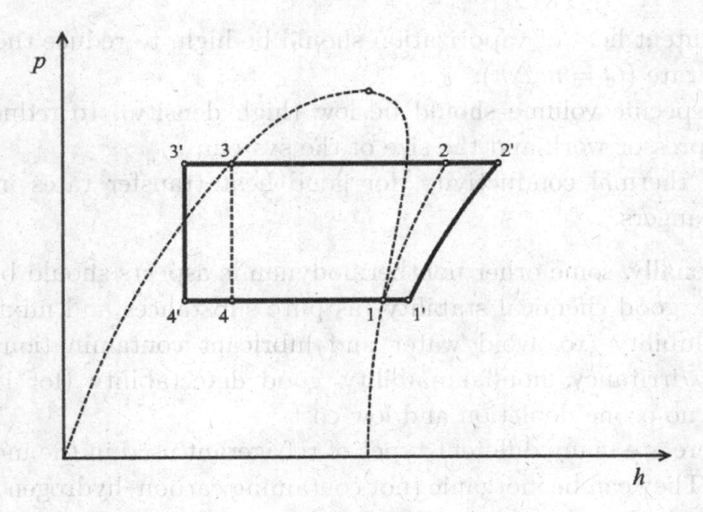

Fig. 10.7

Real processes involving phase transitions cannot stop exactly at the saturation line (see Fig. 10.7). Therefore, some subcooling would be expected at the condenser outlet (point 3′). Subcooled liquid prevents liquid flashing as the refrigerant experiences pressure losses in the tubing and components. Also, because liquid cannot be present in the compressor, some amount of superheat before the compressor inlet (point 1′) becomes necessary to insure the safety of the compressor. Therefore, a more realistic representation of the cycle (neglecting pressure losses) would be 1′2′3′4′.

To be suitable for use in a refrigeration cycle, a refrigerant has to have some particular properties:

- the critical temperature has to be higher than the condenser temperature, to approach the Carnot cycle and hence to achieve a high *COP*;
- the freezing temperature should be lower than the temperature in the evaporator, such that the fluid is kept in liquid state in the evaporator;
- the saturation pressures to be above the atmospheric pressure, to avoid air leaks into the system;

- the latent heat of vaporization should be high, to reduce the mass flow rate ($\dot{Q} = \dot{m}\,\Delta h$);
- the specific volume should be low (high density), to reduce the compressor work and the size of the system;
- high thermal conductivity, for good heat transfer rates in heat exchangers.

Additionally, some other nonthermodynamic aspects should be considered: good chemical stability (as pure substances and mixtures), low solubility (to avoid water and lubricant contamination), low toxicity/irritancy, nonflammability, good detectability (for tracing leaks), no ozone depletion and low cost.

There are many different types of refrigerant used in the industry today. They can be inorganic (not containing carbon–hydrogen bond) or organic in nature. An identifying number is assigned to each refrigerant. It consists of the letter R (for refrigerant) and a suffix made up of digits [44]. For example: R22, R134a, R600a, R717.

10.4. Refrigeration systems

The main reason for using a refrigerator is to keep organic material (e.g., food) cold. Cold temperatures (around +4 °C) slow down the activity of bacteria so that it takes longer for the bacteria to spoil organic matter (food). By freezing it, one can stop the bacteria altogether. Refrigeration and freezing are two of the most common ways to preserve food used today.

Refrigerators are also part of air conditioning systems.

For the ideal vapor-compression refrigeration cycle, the COP is

$$\beta = \frac{\dot{Q}_{\text{in}}}{|\dot{W}_c|} = \frac{\frac{\dot{Q}_{\text{in}}}{\dot{m}}}{|\frac{\dot{W}_c}{\dot{m}}|} = \frac{h_1 - h_4}{h_2 - h_1} = \frac{h_1 - h_3}{h_2 - h_1} \tag{10.8}$$

10.5. Heat pump systems

For geographic zones with moderate heating and cooling needs, heat pumps represent an energy-efficient alternative to furnaces and air conditioners. The most common type of heat pump is the air-source heat pump, which transfers heat between the interior of the house and the outside air. During the heating season, a heat pump moves heat from the cooler outdoors into the warmer house. (During the cooling season, heat pumps act as refrigerators, moving heat from the cooler house into the warmer exterior.) Because they move heat rather than generate heat, heat pumps can provide equivalent space heating at as little as one quarter of the cost of operating conventional heating appliances [45]. As the outside air temperature changes, so does the COP of the heat pump. In northern latitudes, where the annual average temperature is low, the heat pump is found to be uneconomical. It is estimated that, all things considered, the use of a heat pump is not justified when the outside temperature drops below −1 °C [46]. To overcome the problem of temperature variations of air, well water and the earth itself can be used; ground water and the ground itself have a typical minimum temperature of 4–10 °C [46].

For a heat pump, the COP of the ideal cycle is

$$\gamma = \frac{|\dot{Q}_{\text{out}}|}{|\dot{W}_c|} = \frac{|\frac{\dot{Q}_{\text{out}}}{\dot{m}}|}{|\frac{\dot{W}_c}{\dot{m}}|} = \frac{h_2 - h_3}{h_2 - h_1} \tag{10.9}$$

Refrigeration cycles — problems

10.1. A refrigerator using R134a operates in a Carnot refrigeration cycle between $t_C = -15$ °C and $t_H = 30$ °C. Determine the coefficient of performance, the amount of heat absorbed from the refrigerated space, and the net work input.

10.2. Determine the recommended evaporator and condenser pressures for a heat pump that operates on the ideal vapor-compression cycle with refrigerant R134a, if the outside temperature is 5 °C and the inside temperature is to be maintained at 24 °C.

10.3. A heat pump operates on the ideal vapor-compression cycle with refrigerant R134a. The evaporator pressure is 320 kPa and the condenser pressure is 1400 kPa. Determine the power input for a mass flow rate of refrigerant of 0.1 kg/s.

10.4. A refrigerator operates on an ideal vapor-compression refrigeration cycle using R134a as the working fluid between 130 kPa and 750 kPa. The mass flow rate of the refrigerant is 0.06 kg/s. Show the cycle on a T–s diagram and determine: (a) the rate of heat removal from the refrigerated space and the power input to the compressor; (b) the rate of heat rejection to the environment; (c) the coefficient of performance.

Chapter 11

Thermodynamics of Gas Flow

In this chapter, we will deal with flow that involves significant changes in density (compressibility). *Compressible flow* combines fluid dynamics and thermodynamics. We will establish some general relations associated with one-dimensional compressible flow of an ideal gas.

11.1. Velocity of sound

An important parameter in the study of compressible flow is the **velocity of sound** (or the **speed of sound**, or the **sonic speed**) (see [7, p. 286]). This is the speed at which a small pressure disturbance propagates through a fluid and is denoted by symbol c:

$$c^2 = \left(\frac{\partial p}{\partial \rho}\right)_s \tag{11.1}$$

The derivation of Eq. (11.1) can be found in [7]. The bracketed term is the rate of change of pressure with density at constant entropy. It is conventional to use the density ρ in place of the specific volume v in fluid dynamics. It can be shown [47] that when the fluid is an ideal gas ($p = \rho RT$), the differential expression can be written as

$$c = \sqrt{k\,RT} \tag{11.2}$$

where k is the specific heat ratio of the fluid. Equation (11.2) shows that the speed of sound in a specified ideal gas is a function of temperature alone [11, p. 854].

Another important parameter is the **Mach number**,[a] Ma. It is the ratio of the actual velocity of the fluid V and the local sonic velocity c (local refers to the same fluid and same state)

$$Ma = \frac{V}{c} \tag{11.3}$$

The Mach number has no dimension and is often used to describe fluid flow regimes, as follows: subsonic (for $Ma < 1$), sonic (for $Ma = 1$), supersonic (for $Ma > 1$), hypersonic (for $Ma \gg 1$), transonic (for $Ma \approx 1$).

11.2. One-dimensional isentropic flow

The simplest type of flow conceivable is the *one-dimensional* isentropic steady flow. The flow through a duct is said one-dimensional if the following conditions are satisfied [1, p. 419]:

(a) changes in cross-sectional area and curvature of the axis are gradual (Fig. 11.1);
(b) thermodynamic and mechanical properties are uniform across planes normal to the axis of the duct.

No real flow is absolutely one-dimensional, but if changes in direction and area are gradual and if average values of properties at any cross-section are used, the one-dimensional approach yields

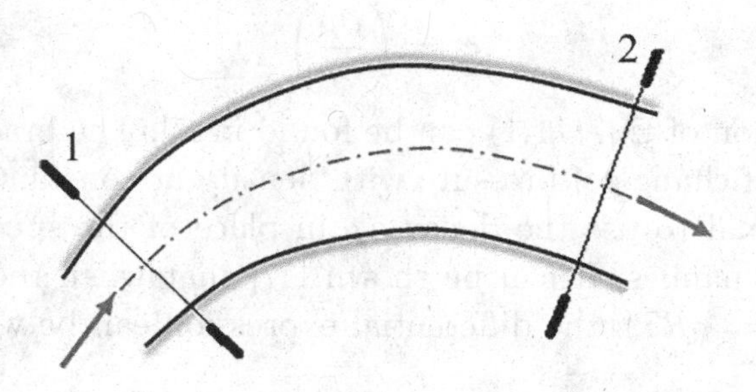

Fig. 11.1

[a]Named after Ernst Mach (1838–1916), an Austrian physicist and philosopher.

sufficiently accurate results for the analysis of compressible flow in nozzles, diffusers and turbine blade passages.

If the flow is isentropic, $\delta q = 0$ and frictional forces are absent. If the flow is steady, the mass flow rate is constant.

Applying the energy equation to the flow between any two planes 1 and 2, given p_1, T_1, and p_2, considering $s_1 = s_2$, given also the initial velocity V_1, the corresponding values of V_2 can be determined.

From the continuity equation ($\dot{m}_1 = \dot{m}_2$), denoting by A is the cross-sectional area normal to the streamline:

$$\frac{A_2 V_2}{v_2} = \frac{A_1 V_1}{v_1} \tag{11.4}$$

The variation of v_2, V_2 and A_2 can be represented by curves of the form shown in Fig. 11.2. All gases and vapors exhibit the same general behavior under these conditions.

Moving from left to right in the diagram in Fig. 11.2, one can draw the following conclusions:

- Specific volume increases continuously. The change in specific volume v is very gradual at low velocities. Therefore, in this case

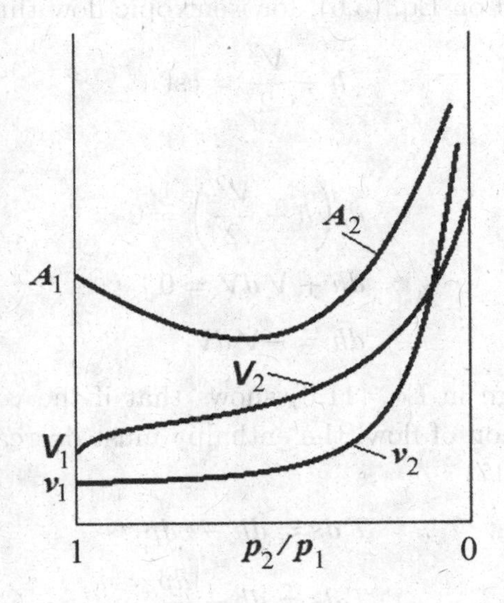

Fig. 11.2

one can consider that the specific volume is constant (incompressible flow). Indeed, as a rule of thumb, flow can be considered incompressible when $Ma < 0.3$. It can be proven that for gas flows with Mach number less than 0.3 the change in density with respect to the change in pressure is less than 5%.

- Velocity increases continuously; its rate of increase is first greater than and finally less than the rate of increase of specific volume (this is typical for gases and vapors).
- Area decreases and then starts to increase; the pressure ratio at which the minimum area is reached is called the **critical pressure ratio**; at this point the mass flow rate per unit area reaches a maximum [1, p. 420].

11.3. Isentropic flow through nozzles

From the steady-flow energy equation for a unit mass of fluid:

$$q - w = \Delta\left(h + \frac{V^2}{2}\right) \tag{11.5}$$

For an isentropic flow where no work is involved, $q = 0$ and $w = 0$. Therefore, based on Eq. (5.6), for isentropic flow through a nozzle

$$h + \frac{V^2}{2} = \text{cst}$$

Differentiating,

$$d\left(h + \frac{V^2}{2}\right) = 0$$

$$dh + V\,dV = 0$$

$$dh = -V\,dV \tag{11.6}$$

The negative sign in Eq. (11.6) shows that if the velocity increases along the direction of flow, the enthalpy must decrease.

From Eq. (6.17)

$$T\,ds = dh - vdp$$

$$T\,ds = dh - \frac{dp}{\rho}$$

For isentropic flow ($ds = 0$),

$$dh = \frac{dp}{\rho} \tag{11.7}$$

Any thermodynamic property can be expressed in terms of two independent intensive properties. Therefore one can write that

$$p = p(\rho, s)$$
$$dp = \left(\frac{\partial p}{\partial \rho}\right)_s d\rho + \left(\frac{\partial p}{\partial s}\right)_\rho ds$$

Since the last term is zero,

$$dp = \left(\frac{\partial p}{\partial \rho}\right)_s d\rho$$

But

$$c = \sqrt{\left(\frac{\partial p}{\partial \rho}\right)_s}$$

therefore

$$dp = c^2\, d\rho \tag{11.8}$$

Combining (11.6) and (11.7),

$$\frac{1}{\rho} dp = -V\, dV \tag{11.9}$$

Because $\dot{m} = \text{cst}$, it follows that $\rho A V = \text{cst}$. Differentiating, $d(\rho A V) = 0$, or

$$AV\, d\rho + \rho V\, dA + \rho A\, dV = 0$$

After dividing by $\rho A V$, one obtains

$$\frac{d\rho}{\rho} + \frac{dA}{A} + \frac{dV}{V} = 0 \tag{11.10}$$

From (11.8), $d\rho = \frac{dp}{c^2}$. From (11.9), $\frac{1}{\rho} = -\frac{V\,dV}{dp}$. Introducing these two equations in (11.10),

$$\frac{dA}{A} = -\frac{d\rho}{\rho} - \frac{dV}{V} = 0$$

$$\frac{dA}{A} = \frac{V\,dV}{c^2} - \frac{dV}{V} = \frac{V\,dV}{c^2} - \frac{V\,dV}{V^2}$$

$$\frac{dA}{A} = V\,dV\left(\frac{1}{c^2} - \frac{1}{V^2}\right) = -V\,dV\left(\frac{1}{V^2} - \frac{1}{c^2}\right) \quad (11.11)$$

$$\frac{dA}{A} = -\frac{V\,dV}{V^2}\left(\frac{V^2}{V^2} - \frac{V^2}{c^2}\right)$$

$$\frac{dA}{A} = -\frac{dV}{V}\left(1 - \frac{V^2}{c^2}\right)$$

In Eq. (11.8) c is the local velocity of sound in the fluid, at the section in the flow where pressure is p, density is ρ and cross-sectional area is A.

Using Ma, Eq. (11.11) becomes

$$\frac{dA}{A} = -\frac{dV}{V}(1 - Ma^2) \quad (11.12)$$

Equation (11.12) may be interpreted without integration, leading to the following important conclusions:

- *Accelerated flow (nozzle)*; dV must be positive so that:
  - When $V < c$, dA is negative, therefore the duct must converge.
  - When $V > c$, dA is positive, therefore the duct must diverge.
- *Decelerated flow (diffuser)*; dV must be negative so that:
  - When $V < c$, dA is positive, therefore the duct must diverge.
  - When $V > c$, dA is negative, therefore the duct must converge.
- *Constant velocity flow*; dV is zero, therefore dA is zero at all times.

It follows immediately that if the flow is to be *continuously* accelerated from a subsonic to a supersonic velocity, the duct must have a *throat* between the two regions, at which $dA = 0$ and the velocity V is equal to the local sonic velocity c [1, p. 422]. Such a duct is called

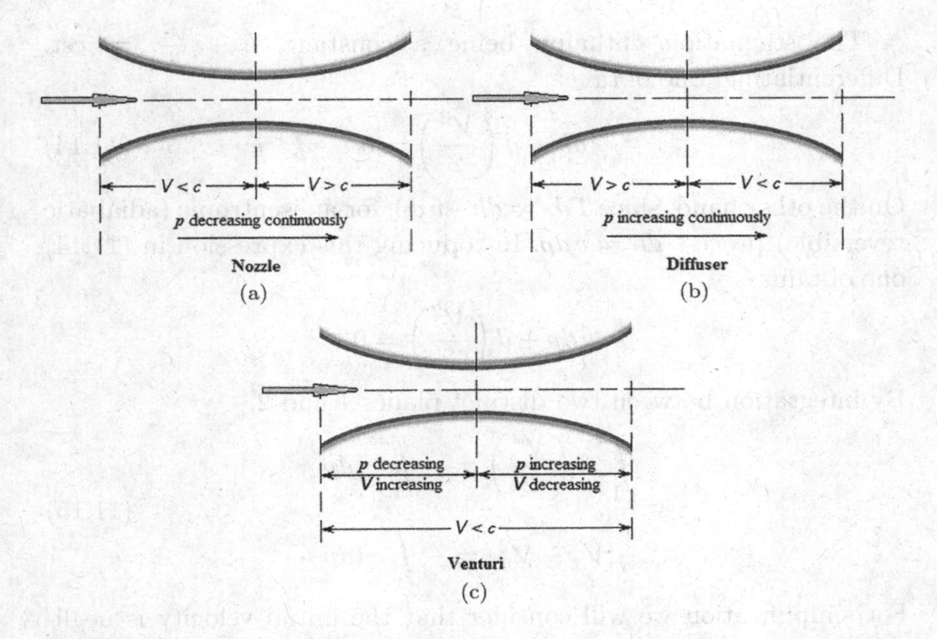

Fig. 11.3

converging–diverging nozzle, also known as *Laval nozzle*.[b] The same remarks apply to the case of a compressive flow continuously decelerated from a supersonic to a subsonic speed. However, when there is no continuous increase (or decrease) of velocity along the duct, the velocity in the throat is *not* sonic. This is the case of a Venturi tube or simply a *venturi*[c] (see Fig. 11.3).

Equation (11.12) can be written also as

$$\frac{dA}{A} = \frac{v\,dp}{V^2}(1 - Ma^2) \tag{11.13}$$

It is worth noting also that the *critical pressure ratio* is that pressure ratio that will accelerate the flow to the local velocity of sound in the fluid.

---

[b]Named after the Swedish engineer and inventor Carl Gustaf Patrik de Laval (1845–1913) who designed it and used it in a steam turbine (1893).
[c]Named after Giovanni Battista Venturi (1746–1822), Italian physicist, man of letters, diplomat and historian of science.

The stagnation enthalpy being a constant, $h + \frac{V^2}{2} = \text{cst}$. Differentiating, one obtains

$$dh + d\left(\frac{V^2}{2}\right) = 0 \tag{11.14}$$

On the other hand, since $T ds = dh - v\,dp$, for an isentropic (adiabatic reversible) process $dh = v\,dp$. Introducing this expression in (11.14) one obtains

$$v\,dp + d\left(\frac{V^2}{2}\right) = 0$$

By integration between two distinct planes 1 and 2,

$$\begin{aligned} \int_1^2 d\left(\frac{V^2}{2}\right) &= -\int_1^2 v\,dp \\ \frac{1}{2}(V_2^2 - V_1^2) &= -\int_1^2 v\,dp \end{aligned} \tag{11.15}$$

For simplification we will consider that the initial velocity is negligible ($V_1 = 0$). For an adiabatic process, ($pv^k = \text{cst}$), Eq. (11.15) becomes

$$\frac{1}{2}V_2^2 = -\int_1^2 \frac{p_1^{1/k}\,v_1}{p^{1/k}}\,dp$$

$$\frac{1}{2}V_2^2 = \frac{k}{1-k}(p_2 v_2 - p_1 v_1)$$

Hence

$$V_2 = \sqrt{\frac{2k}{1-k}p_1 v_1\left[\frac{p_2 v_2}{p_1 v_1} - 1\right]}$$

or

$$V_2 = \sqrt{\frac{2k}{1-k}p_1 v_1\left[\left(\frac{p_2}{p_1}\right)^{(k-1)/k} - 1\right]} \tag{11.16}$$

The maximum possible velocity is obtained when the fluid expands in vacuum ($p_2 = 0$). From Eq. (11.16), the maximum velocity is

$$V_{\max} = \sqrt{\frac{2k}{k-1}p_1 v_1} \tag{11.17}$$

Knowing the state in the exit section 2, using the continuity equation, the mass flow rate $\dot{m}$ can be determined

$$\dot{m} = \frac{A_2 V_2}{v_2} \text{ or } \frac{\dot{m}}{A_2} = \frac{V_2}{v_2} = \frac{V_2 (p_2/p_1)^{1/k}}{v_1} \tag{11.18}$$

An expression of the mass flow rate per unit area can be found by substituting V_2 from (11.16):

$$\frac{\dot{m}}{A_2} = \sqrt{\frac{2k}{1-k}\frac{p_1}{v_1}\left[\left(\frac{p_2}{p_1}\right)^{(k+1)/k} - \left(\frac{p_2}{p_1}\right)^{2/k}\right]} \tag{11.19}$$

This mass flow rate per unit area depends on the pressure ratio p_2/p_1 and is maximum at the throat. To find the value of p_2/p_1 that makes the mass flow rate a maximum, we note

$$\frac{d\left(\frac{\dot{m}}{A_2}\right)}{d\left(\frac{p_2}{p_1}\right)} = 0$$

The solution of this equation is $\frac{p_2}{p_1} = (\frac{2}{k+1})^{k/(k-1)}$. In this equation, p_2 is the throat pressure known as the **critical pressure** p_{cr}. The critical pressure ratio (see also Sec. 11.2) is then

$$\left(\frac{p_2}{p_1}\right)_{cr} = \frac{p_{cr}}{p_1} = \left(\frac{2}{k+1}\right)^{k/(k-1)} \tag{11.20}$$

For some notable values of k, the critical pressure ratios are [1, p. 424]:

- For steam: $k = 1.135$; $(p_2/p_1)_{cr} = 0.577$
- For air: $k = 1.400$; $(p_2/p_1)_{cr} = 0.528$
- For monatomic gases: $k = 1.667$; $(p_2/p_1)_{cr} = 0.487$

From the values listed above one can see that for any gas the critical ratio is around 0.5. For the case of a convergent nozzle, if the pressure ratio $p_2/p_1 > (p_2/p_1)_{cr}$ the exit velocity will be subsonic; if $p_2/p_1 = (p_2/p_1)_{cr}$ the velocity at the exit will be sonic; if $p_2/p_1 < (p_2/p_1)_{cr}$ the exit velocity will also be sonic, the expansion from p_{cr} to p_2 occurring outside the nozzle.

The value of the maximum mass flow rate per unit area can be found by substituting (11.20) in (11.19), obtaining

$$\frac{\dot{m}}{A_{\mathrm{throat}}} = \sqrt{k\left(\frac{2}{k+1}\right)^{(k+1)/(k-1)}\frac{p_1}{v_1}} \tag{11.21}$$

The throat velocity (the critical velocity V_{cr} equal to the local speed of sound) can be found by substituting (11.20) in (11.16):

$$V_{\mathrm{cr}} = \sqrt{\frac{2k}{1+k}p_1 v_1} \tag{11.22}$$

In pneumatics, when a gas leakage rate from a small hole in a pressurized vessel needs to be estimated, it is considered that the hole represents the circular exit section of a convergent nozzle. Knowing the approximate diameter and the pressure ratio, one can estimate the mass flow rate using (11.19) or (11.21).

An interesting engineering application of the Venturi tube is the *ejector*. An air (or steam) ejector is a device which uses the motion of a moving fluid (air or steam) to transport another fluid (gas or liquid). When a Venturi tube is introduced in the pneumatic circuit, it creates a reduced static pressure at the throat, where the velocity of the fluid reaches a local maximum (Venturi effect); because of the vacuum created in the throat (see Fig. 11.3(c)), if a suction line is provided (Fig. 11.4), a secondary fluid (*suction fluid*) can be absorbed and mixed with the main fluid (*motive fluid*). Ejectors can be used in a variety of applications where a high pressure motive fluid absorbs and compresses low pressure suction fluid to discharge it at an intermediate pressure.

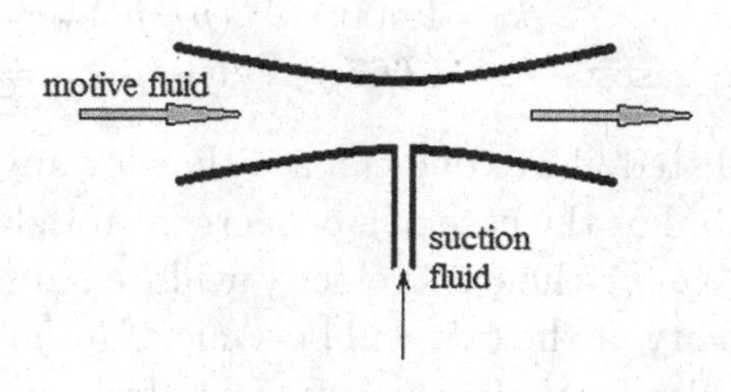

Fig. 11.4

Thermodynamics of gas flow — problems

11.1. Air behaving like a perfect gas enters a nozzle at 200 kPa and 150 °C with zero initial velocity (standstill) and leaves the nozzle at 100 kPa. The exit area of the nozzle is 0.03 m^2. Assuming steady isentropic flow, determine the type of nozzle, the exit velocity and the mass flow rate of air.

11.2. At some location inside a nozzle, the air temperature is 423 K and the velocity is 460 m/s. What kind of flow regime exists at that point?

11.3. Air from a large tank at $p_1 = 200$ kPa and $T_1 = 350$ K is supplied to a convergent–divergent nozzle. What is the temperature in the adiabatic nozzle where the Mach number is $Ma = 1.2$?

11.4. A pressure vessel contains N_2 gas pressurized at 2.5 bar at $t_1 = 20$ °C. Estimate the leakage rate when a small hole of circular shape with diameter $d = 0.1$ mm develops, if the discharge coefficient is $c_d = 0.67$.

Chapter 12

Combustion Processes

12.1. Fuels and combustion

In many industrial processes, the primary source of energy is heat produced by combustion of fuels. This is the case of IC engines, steam power plants and gas power plants, to name just a few. **Combustion** is a chemical reaction between a *fuel* and an *oxidizer* accompanied by the release of a large quantity of thermal energy. A **fuel** is any material that can be burned to release thermal energy. An **oxidizer** (also called *oxidant*) is a chemical compound that readily transfers oxygen atoms. The oxidizer most often used in combustion processes is oxygen from the air. However, chemical substances such as nitrates, chlorates and peroxides can also act as oxidants.

The most common materials used in industrial combustion processes are the so-called **fossil fuels**. They are considered (biological hypothesis) as remains of dead plants and animals fossilized by exposure to heat and pressure in the Earth's crust over hundreds of millions of years. Fossil fuels are coal, petroleum (fuel oil) and natural gas or products derived from them.

Fossil fuels are still the dominant resource, providing 80 % of energy, while new renewables (solar, wind, geothermal, marine) provide about 1.5 % only. The share of coal is around 28 %, oil has 31 %, while natural gas has 23 %. For electricity production, fossil fuels supply 66 %, while new renewables supply around 5 %. [48]. According to the World Coal Association [49], coal provides around

30 % of global primary energy needs and generates over 40 % of the world's electricity.

Fuel oil and natural gas are hydrocarbon fuels (C_mH_n). Combustible elements are: carbon (C), hydrogen (H_2) and sulfur (S_2). Nitrogen (N_2) oxidizes only in special cases.

In general in a combustion reaction,

$$\text{Reactant} + \text{Oxygen} \rightarrow \text{Oxide(s)} + \text{heat}$$

Examples of oxidation reactions are:

Carbon to (CO)	$2C + O_2 \rightarrow 2CO$
Carbon to (CO_2)	$C + O_2 \rightarrow CO_2$
Carbon Monoxide	$2CO + O_2 \rightarrow 2CO_2$
Hydrogen	$2H_2 + O_2 \rightarrow 2H_2O$
Sulfur to (SO_2)	$S + O_2 \rightarrow SO_2$
Sulfur to (SO_3)	$2S + 3O_2 \rightarrow 2SO_3$
Methane	$CH_4 + 2O_2 \rightarrow CO_2 + 2H_2O$
Acetylene	$2C_2H_2 + 5O_2 \rightarrow 4CO_2 + 2H_2O$
Ethylene	$2C_2H_6 + 7O_2 \rightarrow 4CO_2 + 6H_2O$
In general, for hydrocarbons:	$C_mH_n + (m + n/4)\ O_2 \rightarrow m\,CO_2 + n/2\ H_2O$

Next, some important concepts related to combustion and fuels will be briefly introduced.

The **ignition temperature** is the lowest temperature at which combustion spontaneously begins and continues in a substance when it is heated in normal atmosphere without an external source of ignition such as a flame or spark. It is also known as *autoignition temperature* or *ignition point.* The autoignition temperature for a given fuel decreases as the pressure increases or as the oxygen concentration increases.

The combustion process can be *complete* — if all the carbon in the fuel burns to CO_2, all the hydrogen to H_2O and all the sulfur to SO_2 — or *incomplete* — if carbon burns to CO, or combustion products contain unburned elements such as C, H_2, etc.

Stoichiometric or **theoretical combustion** is the ideal combustion process where fuel is burned completely. The minimum amount of air necessary to burn completely a given amount of fuel is called *stoichiometric air* or *theoretical air*. It can be calculated from the chemical reaction.

A stable combustion requires the right amounts of fuel and oxygen. Because burning conditions are never ideal, more air than ideal must be supplied to burn all fuel completely in industrial burning devices. Thus, power plant boilers normally run about 10–20 % **excess air**. Natural gas-fired boilers in power plants may run as low as 5 % excess air (1–2 % in some cases). Pulverized coal-fired boilers may run with 20 % excess air [50].

The **heating value** is the amount of heat released when a fuel is burned completely and the products are returned to the state of the reactants. Heating values are expressed in thermal energy units per unit mass (J/kg, Btu/ lb) for solid and liquid fuels, and in thermal energy units per unit volume at *standard conditions* (J/Sm^3, Btu/Sft^3).

There are two different methods to measure the amount energy released and, consequently, two types of heating values:

- *LHV* — lower heating value, obtained under such conditions that water in the products is in vapor state (also called *net calorific value, NCV*), and
- *HHV* — higher heating value, when water in the products is in liquid state (also called *gross calorific value, GCV*).

When burning fossil fuels in industrial burning devices, only the *LHV* can be utilized. For coal and oil, the difference between *LHV*s and *HHV*s is approximately 5 %. For most natural gas and manufactured gases, the difference is approximately 9–10 % [51].

For reference, the average *LHV* of coal is about 29,000 kJ/kg and that of oil about 42,000 kJ/kg.

12.2. Adiabatic flame temperature

If the entire heat produced through combustion is stored in the combustion products (no heat losses, $Q = 0$), the temperature of the products is maximum and it is called the **adiabatic flame temperature** or *adiabatic combustion temperature* [11, p. 792].

For a certain fuel, this temperature depends on the state of the reactants, the degree of completion of the reaction and the amount of air used. For a given fuel at a specified state, the maximum adiabatic flame temperature occurs with the theoretical amount of air. For coal, this is over 1000 K; for hydrocarbons, about 2000 K.

12.3. Effectiveness of energy conversion

For a combustion process, the chemical energy stored in a system is converted into thermal energy. The performance index for this process is called **combustion efficiency**. Using the standard relationship,

$$\eta_{\text{combustion}} = \frac{Q}{HV} = \frac{\text{heat released during combustion}}{\text{heating value of the fuel burned}} \quad (12.1)$$

The definition of efficiency should be clear with respect to what type of heating value is used. In the United States and Canada, the standard is to use the HHV; however, engineers and manufacturers dealing with gas turbines normally use LHV when quoting heat rates or efficiencies. In other countries, LHV is used for all specifications of fuel consumption and unit efficiency [52].

General estimations [53] show that even the most efficient fossil fuel power stations lose at least half of their energy in conversion. The internal combustion engine also represents a very inefficient way of converting primary energy to movement.

One of the reasons for such low efficiencies is the fact that any power system includes at least two energy conversion devices and the efficiency of a system is equal to the product of efficiencies of the individual devices (subsystems). When one of the devices in the system converts thermal energy into work, it will be affected by the Carnot efficiency, which is never too high. Figure 12.1 shows the case of a thermal power plant.

An alternative to that is the use of **fuel cells**. A fuel cell is an electrochemical device that converts directly the chemical energy of a fuel (hydrogen, hydrocarbons, alcohol, etc.) into electricity (Fig. 12.1) and therefore its efficiency is not limited by the Carnot efficiency. The concept is not new, it was discovered in 1838 by the

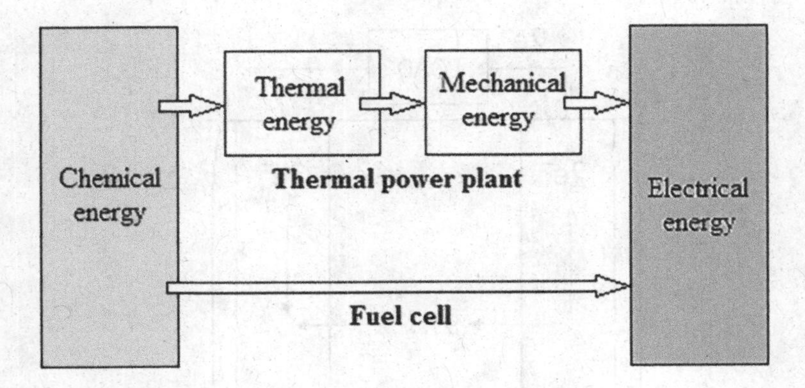

Fig. 12.1

German-Swiss chemist Christian F. Schönbein. In principle, a fuel cell operates based on a reverse water electrolysis reaction: if the power source is removed, a fuel is fed to the anode to supply hydrogen while oxidant is fed to the cathode to supply oxygen, what results is a potential difference between the two electrodes. The device becomes an electric generator that will operate for as long as fuel and oxidant are supplied. A typical fuel cell (see Fig. 12.2) consists therefore of two *electrodes* (*anode* and *cathode*) separated by an *electrolyte* — a material that blocks the electrons but allows ions to circulate. The fuel (*hydrogen* based) is fed to the anode. H_2 will turn into H^+ ions and will release electrons which will be captured by the anode. The H^+ ions reach the cathode (through the electrolyte) where they combine with the *oxygen* from the air supplied, to produce water. The transfer of electrons toward the cathode through the external circuit will produce a direct current, whereas in the electrolyte the current transfer is produced by transfer of ions.

The anode and cathode may be porous to allow distribution of oxygen or hydrogen. Because each individual cell does not generate more than about 0.7 V DC, a large number of such cells have to be connected in series to provide a useful power output. Sometimes it is necessary to introduce in the circuit a DC/AC converter. Different types of fuel cells operate at different temperatures, from under 100 °C to over 1000 °C. In fuel cells that operate at low temperatures,

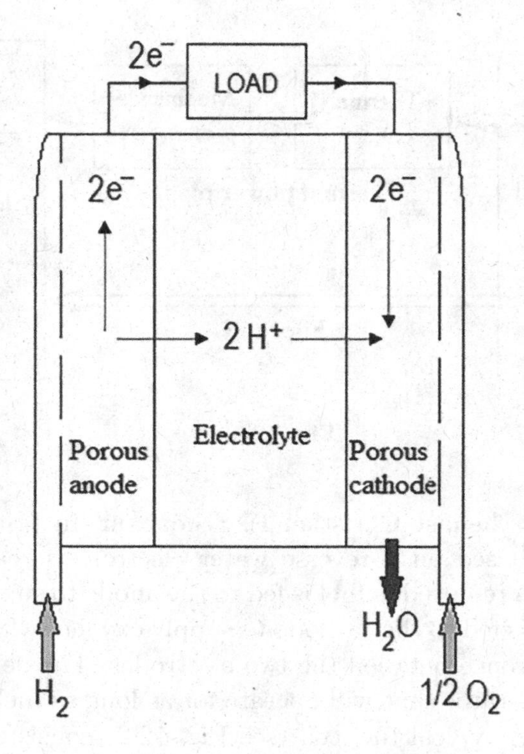

Fig. 12.2

the reverse electrolysis process is accelerated by means of a *catalyst*: in general, this is a fine layer of platinum applied over the electrodes.

Although fuel cells represent a big improvement over IC engines, coal burning power plants, and nuclear power plants (all of which produce harmful byproducts), currently hydrogen is produced mostly by reforming of fossil fuel with massive emission of CO_2, some sulfur dioxide, and nitrous oxides. Nonetheless, fuel cells are relatively nonpolluting, quiet, easy to maintain (no moving parts), with conversion efficiencies of about 50 %.

In a higher temperature range, the electrodes may also have channels to allow the distribution of coolants, such as water. The waste heat provided by fuel cells that operate at high temperatures can be used for heating. Such "cogenerating" systems can approach overall efficiencies of up to 80 % in ideal circumstances [54].

One major disadvantage of fuel cells is the high cost driven by the platinum catalyst. Fuelling fuel cells represents another problem since the production, transportation, distribution and storage of hydrogen are difficult.

An exciting application of fuel cells is for automotive propulsion [55, 56] although it appears that these propulsion systems have prohibitive costs, making them unlikely to become competitive in the very near future, despite of their obvious vehicle emissions benefits. Still, the Japanese automaker Toyota unveiled, at the November 2014 Los Angeles Auto Show, its new hydrogen fuel cell vehicle, Toyota Mirai. By January 2018, the company sold more than 3000 units in California. The 2014 price including tax was over € 78,000 (in Germany [57]). In 2017, the 153 HP Mirai, with a full-fuelled range of about 480 km, sold for about US$ 57,500 (MSRP).

Combustion processes — problems

12.1. What fuels are used for industrial boilers, in general? (a) coal; (b) propane; (c) solid, liquid or gas.

12.2. *LHV* stands for: (a) lower standard value; (b) large heating value; (c) lower heating value.

12.3. The adiabatic flame temperature ______ when pure oxygen instead of air is fed.
(a) increases; (b) decreases; (c) remains constant.

12.4. The main reason(s) why the car industry is developing fuel cells is/are: (a) no CO_2 emissions; (b) engine about 10 times lighter; (c) alternative to non-renewable fossil fuels.

Chapter 13

Heat Transfer

According to the second law of thermodynamics, heat "flows", spontaneously, only from a high to a low temperature system. The "driving force" of heat transfer is, therefore, the difference in temperature between systems, or a system and its surroundings. For this reason, in what follows we are going to waive the sign convention used before; heat exchanged will be considered positive whether it is transferred *from* or *to* a system. Also, for all cases of heat transfer it will be considered that the temperature at every point within the system, including its surfaces, is independent of time. This allows the use of a *steady-state heat transfer analysis* to determine the temperature distribution and heat flow. The value of the heat transfer rate depends not only on the temperature difference, but also on the thermophysical properties, size, geometric shape and relative movement of the heat exchanging entities [58, p. 6].

There are three mechanisms (modes) by which thermal energy is transferred: *conduction*, *convection* and *radiation*. They may occur separately or in combination.

Conduction is heat transfer by means of molecular agitation within a material without any motion of the material as a whole (no bulk motion).

Convection is heat transfer by bulk (macroscopic) motion of a fluid such as air or water in contact with a surface having a different temperature.

Radiation is heat transfer by the emission of electromagnetic waves which carry energy away from the emitting object.

13.1. Conduction

Conduction is predominant in solids. There the energy is transmitted through lattice vibrations induced by vibrations of atoms about their equilibrium position. In nonmetals the transfer takes place exclusively through lattice vibrations. In metals, the motion of free electrons also contributes to heat conduction, through collisions.

Conduction in gases and liquids is less common than in solids and occurs through random molecular motion (diffusion). The molecules in liquids and gases are spread farther apart, so it takes longer for liquids and gases to conduct heat. Light gases (e.g., H, He) typically are more conductive; dense gases such as Xe, Ar, dichlorodifluoromethane (R-12), are less conductive.

The most general (vector) equation for multidimensional conduction is

$$\vec{\dot{Q}}'' = -k \nabla T \tag{13.1}$$

where $\vec{\dot{Q}}''$ is the *Fourier heat flux vector* and T is the steady temperature field $T = T(x, y, z)$ with the gradient ∇T. It is worth noting that the temperature and the difference in temperature can be measured either in kelvin (T, ΔT) or in degrees Celsius (t, Δt) (see also Sec. 7.2). (In this section $\dot{Q}$ is used for heat transfer rate, $\dot{Q}'$ for heat transfer rate per unit length, $\dot{Q}''$ for heat transfer rate per unit area or heat flux.) This implies that:

- The heat transfer occurs in the direction of decreasing temperature (reason for the "–" sign).
- The direction of heat transfer is perpendicular to the lines of constant temperature (isotherms).

The fundamental equation for conduction in the normal direction n is therefore:

$$\dot{Q}''_n = -k\frac{\partial t}{\partial n} \tag{13.2}$$

If this direction coincides with the x-coordinate, then one can write:

$$\dot{Q}''_x = -k\frac{dt}{dx} \tag{13.3}$$

This is known as the **Fourier's law**[a] **of one-dimensional conduction**, where $\dot{Q}''_x$ is the **heat flux** in [W/m^2], k is the **thermal conductivity** of the material, expressed in [W/(m K)] or [W/(m °C)], and $\frac{dt}{dx}$ is the **temperature gradient** in the x-direction. For a steady-state process, $\dot{Q}''_x$ and consequently $\frac{dt}{dx}$ are constant in time. Thermal conductivity k is a material property; it may vary with temperature and pressure. Some typical values can be found in the Appendix E, Table A.9. In the US Customary System, k is measured in [Btu/(h ft °F)]. Here are the conversion formulas:

$$1\text{W/(m K)} = 0.57779 \text{ Btu/(h ft °F)}$$
$$1\text{Btu/(h ft °F)} = 1.73073 \text{ W/(m K)}$$

The heat flow rate $\dot{Q}_x$ [W] through a defined cross-sectional area A [m^2] perpendicular to the direction of heat flow will be:

$$\dot{Q}_x = \dot{Q}''_x A = -k\,A\frac{dt}{dx} \tag{13.4}$$

13.1.1. *Steady-state conduction through a plane wall*

If the heat transfer occurs through a plane wall (see Fig. 13.1) from surface 1 (where $x = 0$ and $t = t_{s1}$) to surface 2 (where $x = L$ and $t = t_{s2}$), then

$$\dot{Q}_x = -kA\frac{dt}{dx}$$

[a]Named after Jean-Baptiste Joseph Fourier (1768–1830), a French mathematician and physicist.

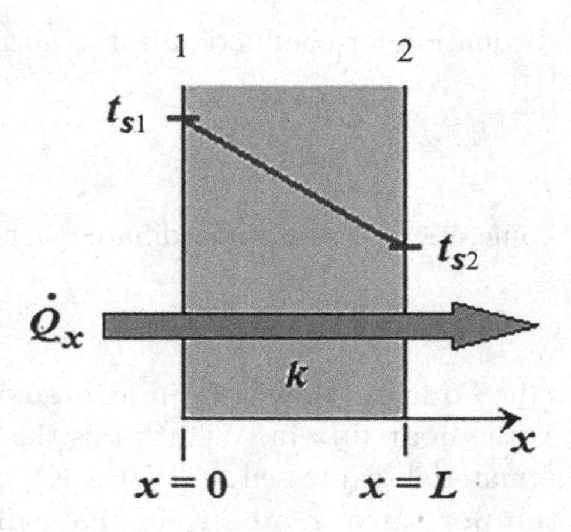

Fig. 13.1

Separating the variables,

$$\dot{Q}_x dx = -kA\,dt$$

Integrating,

$$\int_1^2 \dot{Q}_x dx = \int_1^2 -kA\,dt$$

$$\dot{Q}_x \int_0^L dx = -kA \int_{t_{s1}}^{t_{s2}} dt$$

$$\dot{Q}_x(L-0) = -kA(t_{s2} - t_{s1}) \tag{13.5}$$

$$\dot{Q}_x = -\frac{kA(t_{s2} - t_{s1})}{L}$$

$$= \frac{k}{L}A(t_{s1} - t_{s2}) = \frac{k}{L}A(t_{\text{hot}} - t_{\text{cold}})$$

The temperature distribution

If heat is conducted through a homogeneous plane wall, then $\frac{dt}{dx}$ is nonzero and constant with depth

$$\frac{dt}{dx} = \frac{\Delta t}{\Delta x} = C_1 \tag{13.6}$$

where C_1 is a constant.

From Eq. (13.6),

$$dt = C_1\,dx$$

$$\int dt = C_1 \int dx \tag{13.7}$$

$$t = C_1 x + C_2$$

Considering the boundary conditions:

$$x = 0, \quad t = t_{s1}$$
$$x = L, \quad t = t_{s2}$$

one obtains:

$$C_2 = t_{s1}$$
$$C_1 = \frac{t_{s2} - t_{s1}}{L}$$

Thus, Eq. (13.7) becomes

$$t(x) = (t_{s2} - t_{s1})\frac{x}{L} + t_{s1} \tag{13.8}$$

Equation (13.8) shows that the temperature distribution in a plane wall is linear.

Thermal resistance and the electrical analogy

There is an electrical analogy with conduction heat transfer that can be used in problem solving. Ohm's law can be expressed as

$$I = \frac{\Delta V}{R}$$

where I is the current flowing through an element of circuit, ΔV is the voltage across the element, and R is the electrical resistance across the element. Similarly Fourier's law can be written as

$$\dot{Q}_x = \frac{\Delta t}{R_{t,\mathrm{cond}}} = \frac{\Delta t}{\frac{L}{k\,A}} \tag{13.9}$$

where q_x is the rate of heat conduction in the direction x, Δt is the temperature difference between the surfaces of a plane wall, and $R_{\mathrm{t,cond}}$ is the thermal resistance for conduction between the surfaces defined as

$$R_{t,\mathrm{cond}} = \frac{L}{kA}$$
$$[R_{t,\mathrm{cond}}]_{\mathrm{SI}} = \mathrm{K/W} \text{ or } {}^{\circ}\mathrm{C/W} \tag{13.10}$$
$$[R_{t,\mathrm{cond}}]_{\text{US Customary}} = (\mathrm{h}\ {}^{\circ}\mathrm{F})/\mathrm{Btu}$$

The electrical analogy proves very useful when dealing with composite walls of complex structures. For a plane wall consisting of three layers a, b and c (see Fig. 13.2),

$$R_{t,\mathrm{cond}} = \sum R_i = R_a + R_b + R_c = \frac{L_a}{k_a A} + \frac{L_b}{k_b A} + \frac{L_c}{k_c A} \tag{13.11}$$

Moreover, the electrical analogy is not limited to direct current; concepts from the alternating current theory can be applied to the modeling of thermal transient phenomena.

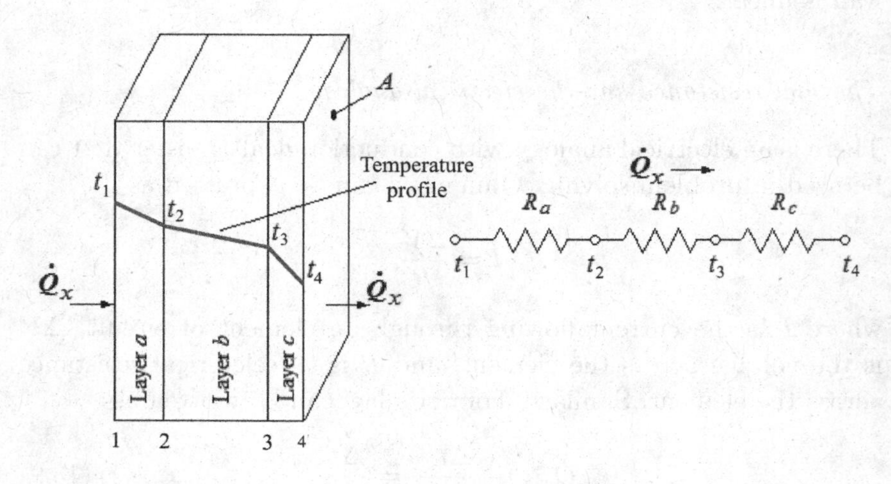

Fig. 13.2

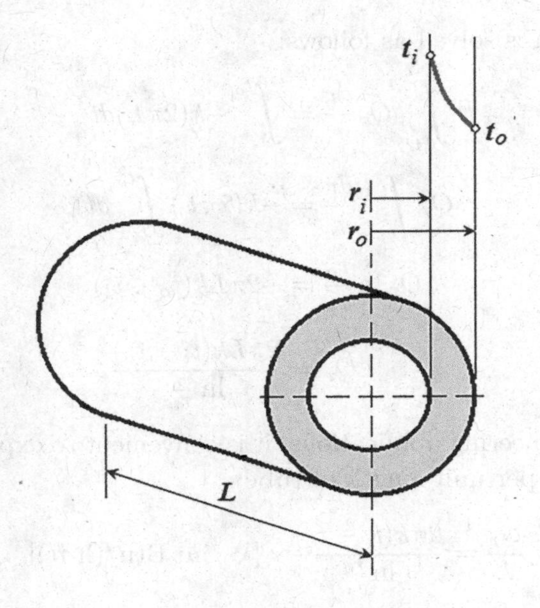

Fig. 13.3

13.1.2. *Steady-state conduction through a cylindrical wall*

Assume a tube (see Fig. 13.3) having the inside surface of radius r_i and an outside surface r_o with inner and outer surface temperature maintained at t_i and t_o, respectively ($t_i > t_o$).

The heat flows in the radial direction with the only independent variable being the radius r.

Equation (13.4) can be rewritten as

$$\dot{Q}_r = -k\frac{dt}{dr}A$$

For a cylinder, $A = 2\pi rL$, where L is the length of the cylindrical wall.

$$\dot{Q}_r = -k(2\pi rL)\frac{dt}{dr}$$

Separating variables,

$$\dot{Q}_r\frac{dr}{r} = -k(2\pi L)dt$$

The equation is solved as follows:

$$\begin{aligned}
\int_{r_i}^{r_o} \dot{Q}_r \frac{dr}{r} &= \int_{t_i}^{t_o} -k(2\pi L)dt \\
\dot{Q}_r \int_{r_i}^{r_o} \frac{dr}{r} &= -k(2\pi L) \int_{t_i}^{t_o} dt \\
\dot{Q}_r \ln \frac{r_o}{r_i} &= -2\pi Lk(t_o - t_i) \\
\dot{Q}_r &= \frac{2\pi Lk(t_i - t_o)}{\ln \frac{r_o}{r_i}}
\end{aligned} \tag{13.12}$$

In many engineering applications, it is convenient to express the heat transfer rate per unit length of tube:

$$\dot{Q}'_r = \frac{\dot{Q}_r}{L} = \frac{2\pi k(t_i - t_o)}{\ln \frac{r_o}{r_i}} \quad \text{[W/m; Btu/(h ft)]} \tag{13.13}$$

Equation (13.12) shows that under steady-state conditions the temperature difference in radial direction is no longer linear, but follows a logarithmic curve. The thermal resistance is

$$\dot{Q}_r = \frac{2\pi\, k\, L(t_i - t_o)}{\ln\left(\frac{r_o}{r_i}\right)} = \frac{(t_i - t_o)}{R_{t,\text{cond}}} \tag{13.14}$$

$$R_{t,\text{cond}} = \frac{\ln\left(\frac{r_o}{r_i}\right)}{2\pi\, k\, L} \tag{13.15}$$

For a multilayer cylindrical wall, based on the electrical analogy:

$$R_{\text{tot}} = \sum R_i \tag{13.16}$$

where i is the layer index.

13.2. Convection

The term *convection* comes from the Latin *convehere* "to carry together" (*com-* "together" + *vehere* "to carry") which suggests the mass motion of the fluid during the heat exchange.

The heat transfer from the solid surface to the fluid can be described by *Newton's law of cooling.* It states that the heat flux, from a solid element of surface temperature t_s to a fluid of bulk temperature t_∞, is

$$\dot{Q}'' = h(t_s - t_\infty) \tag{13.17a}$$

where $\dot{Q}''$ is the convective heat flux [W/m^2], h is the **convection heat transfer coefficient** [W/(m^2 K) or W/(m^2 °C)], t_s is the surface temperature [K or °C], and t_∞ is the fluid bulk temperature [K or °C]. The subscript ∞ is used here to designate conditions in the free stream outside the boundary layer.

In Eq. (13.17a) it is assumed that $t_s > t_\infty$, in which case heat is transferred from the solid surface to the fluid, but the equation can be rewritten for the case $t_\infty > t_s$:

$$\dot{Q}'' = h(t_\infty - t_s) \tag{13.17b}$$

Newton's law of cooling shows that the convective heat flux is proportional to the difference between the surface and fluid temperatures, h acting as a proportionality constant.

For the entire area of contact between the fluid and the solid, the total heat transfer rate $\dot{Q}$ (W) can be expressed as

$$\dot{Q} = \dot{Q}''A_s = hA_s(t_s - t_\infty) \tag{13.18}$$

where A_s is the area of the surface exposed to fluid flow [m^2]. Unlike the thermal conductivity k, h is not a physical constant. It depends upon the fluid properties (density, viscosity, specific heat, thermal conductivity — which are functions of t), the surface geometry, and the flow conditions. Also, h is actually an **average convection coefficient**, because temperatures vary continuously during the heat transfer process. In addition to that, the presence of a boundary layer for heat transfer has to be considered. Consequently, the value of h needs to be determined for each and every system analyzed. This is called *the problem of convection.*

Convection may be of two types depending on the mechanism of flow: *free* (natural) convection and *forced* convection.

In **free convection**, the motion of the fluid is caused only by the differences in density that can be maintained in the fluid due to the temperature differences. For example, when boiling water in a pot, the heat from the burner is transferred to the water at the bottom; then, this hot water rises and cooler water moves down to replace it, causing *convection currents*. Another example is offered by the air heater in a room; the space heater (also known as "radiator") consists of a series of elements through which hot oil (electrically heated) or hot water (in the case of central heating) flows. The heater is placed close to the floor. Based on the ideal gas equation of state, at constant pressure,

$$\rho T = \frac{p}{R} = \text{cst}$$

Therefore if the neighboring air is heated, its density must decrease at the same rate, making it buoyant. As raising air is cooled at the ceiling and outside walls, it becomes denser and eventually sinks to the floor moving toward the space heater. A properly designed heating system using natural convection circulation can be quite efficient in uniformly heating a home.

In **forced convection**, the motion of the fluid is imposed externally, by a pump or fan. For example: a fan-powered heater, where a fan blows cool air past a heating element, heating the air. A person blowing on their food to cool it is using also forced convection.

The fluid flow pattern can be either *laminar* or *turbulent*. The **laminar flow** generally occurs in relatively low velocities, when fluid travels smoothly or in regular paths. It may be thought of as consisting of thin, parallel, layers[b] that slide over each other. If the flow in a pipe is laminar, the distribution of velocity has a parabolic shape with the maximum velocity at the center. **Turbulent flow** develops at higher velocities, with the formation of eddies and chaotic motion. For fully developed turbulent flow, the velocity distribution across a section of the pipe is fairly flat (see Fig. 13.4).

At the interface between fluid and a solid wall, there is always a thin layer δ_t of fluid with a temperature distribution from that of

[b] *Lamina* in Latin means "thin layer of material".

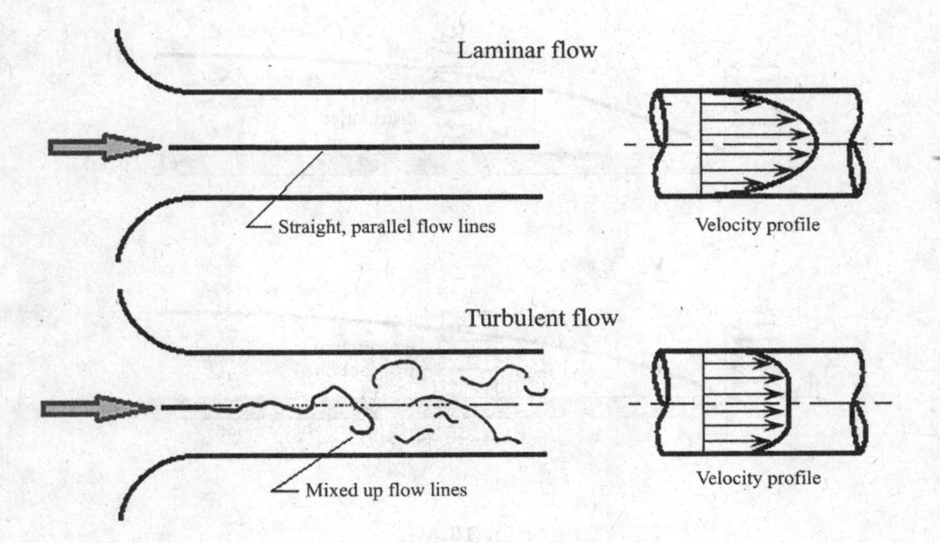

Fig. 13.4

the freestream t_∞ to that of the surface t_s. This is called **thermal boundary layer** [7, p. 344] (see Fig. 13.5(b)). It coexists with the **hydrodynamic** (velocity) **boundary layer** (see Fig. 13.5(a)), characterized by a velocity gradient; in this layer the flow is laminar, regardless of the flow pattern of the rest of the fluid. The presence of the thermal boundary layer increases the complexity of the convective phenomena because through this layer heat is transferred by conduction. However, when there is an appreciable movement of the fluid, conduction heat transfer in fluid may be neglected compared with convection heat transfer.

Because of the large number of independent variables that influence the value of h, the differential equations of flow and heat transfer are complicated and difficult to solve analytically. The solution consists of applying the concept of **similitude** to thermodynamic phenomena, using the method of *dimensional analysis*, to define a number of dimensionless groups using variables that have significance in a general flow and heat transfer field. Two concepts need to be defined first: *model* and *prototype*. From an engineering standpoint, a **model** is a reduced-scale representation of a physical system that can

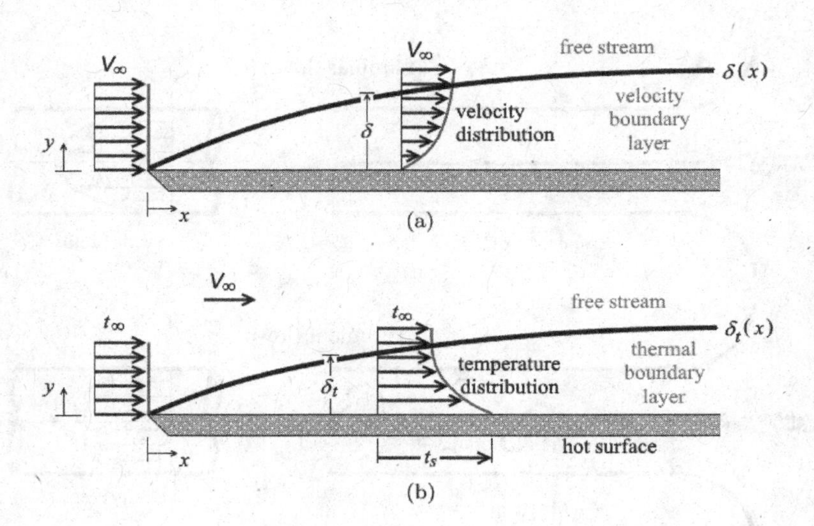

Fig. 13.5

be used to predict a specific behavior of the system. The **prototype** is the actual (full scale) physical system for which the predictions are to be made [7, p. 304]. It can be shown that the model and the prototype will display similar behavior in some desired respect if a specific dimensionless group[c] has the same value for both model and prototype. The dimensionless groups are also called **similitude criteria**. The use of these criteria presents the following important advantages: a significant reduction in the number of variables to be investigated, making the results independent of the scale of the system and of the system of units being used and simplification of the scaling-up or scaling-down of results obtained with models of systems. For convective heat transfer,

$$h = f(\text{similitude criteria}) \tag{13.19}$$

The expression of function f is determined experimentally, based on tests performed on models. The functional relationship f and the similitude criteria involved depend on the type of convection (natural or forced).

[c] An example of dimensionless group is the Mach number, M. More details on how to define relevant dimensionless groups can be found in [7, p. 304].

13.2.1. *The case of forced convection*

It has been determined experimentally that, for forced convection, a most general correlation applicable involving dimensionless groups is

$$Nu = C\, Re^m\, Pr^n \tag{13.20}$$

where Nu is the average Nusselt number, Re is the Reynolds number, Pr is the Prandtl number and C, m and n are numerical constants. The values of C, m and n depend on the flow conditions and are determined experimentally. Solving for Nu and knowing k and L, the value of h can be determined.

It has been determined experimentally that the degree of turbulence is estimated by the **Reynolds number**[d] (Re).

$$Re = \frac{\rho V L}{\mu} = \frac{V L}{\nu} \tag{13.21}$$

where ρ is the density of the fluid [kg/m^3], V is its velocity [m/s], μ is the dynamic viscosity [Pa s or kg/(m s)], ν is the kinematic viscosity [m^2/s], L is a characteristic length of the surface [m]. The choice of L depends on the system geometry. For example, for a flow through or across a circular pipe, L is the pipe diameter (inner or outer, respectively).

The Re number may be interpreted as the ratio of two forces that influence the behavior of fluid flow in the boundary layer: inertia forces and viscous forces (see Appendix B).

Depending on the Re number, three flow regimes are noted for the flow in a tube:

(a) laminar flow: $Re < 2320$
(b) transitional flow: $2320 < Re < 10\,000$
(c) turbulent flow: $Re > 10\,000$

In the absence of upstream disturbances, any fluid flow is laminar at the leading edge of the surface and then may become turbulent under specific conditions. The transition to turbulence may occur at

[d]Named after Osborne Reynolds (1842–1912), British physicist, famous for his contributions to fluid dynamics.

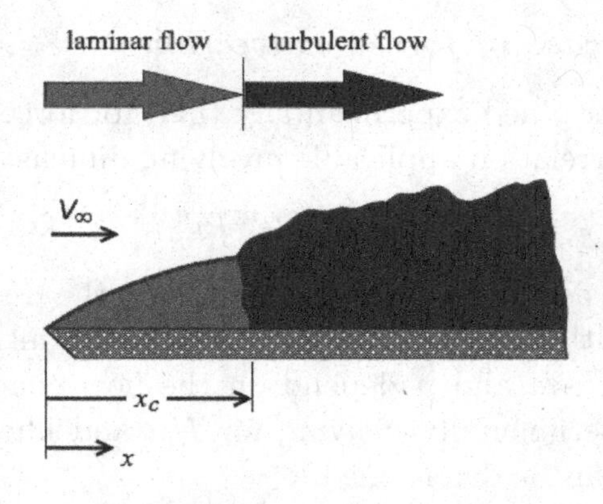

Fig. 13.6

a distance from the leading edge called critical distance x_c, where the value or Re is critical Re_c (Fig. 13.6). The transition from laminar to turbulent flow occurs at a Reynolds number of approximately 2320 in a pipe, so $Re_c = 2320$. If flow occurs under carefully controlled conditions (smooth pipe, no disturbances, etc.), a laminar flow can be obtained for $Re > Re_c$. However, if Re is less than 2320, the flow will always be laminar.

For flow along a plane surface, the following domains are considered:

(a) laminar flow: $Re < 500\,000$
(b) turbulent flow: $Re > 500\,000$

For viscous fluids (fuel oil, oils) flow is usually laminar; less viscous fluids (gases, water, steam) have usually a turbulent flow.

The **Prandtl number**[e] (Pr) is the dimensionless number consisting of a grouping of the properties of the fluid. The Prandtl number is the ratio of momentum diffusivity (kinematic viscosity) to thermal diffusivity. It is defined as

$$Pr = \frac{\nu}{\alpha} = \frac{\mu c_p}{k} \tag{13.22}$$

[e]Named after the German engineer Ludwig Prandtl (1875–1953).

Table 13.1

Material	***Pr***
Liquid metals	0.004–0.03
Gases	0.7–1.0
Water	1.7–13.7
Oils	50–100 000

where ν is the kinematic viscosity [m^2/s], α is the thermal diffusivity [m^2/s], μ is the dynamic viscosity [Pa.s] or [kg/(m s)], c_p is the specific heat [J/(kg K)], and k is the thermal conductivity [J/(m K)].

The Prandtl number is actually the ratio of thicknesses of flow boundary layer δ to thermal boundary layer δ_t. When $Pr = 1$, the boundary layers coincide. If $Pr > 1$, the thermal boundary layer is thinner. Typical values of the Prandtl number are presented in Table 13.1.

The **Nusselt number**[f] (Nu) is the dimensionless number that appears when dealing with convection. It provides a measure of the convection heat transfer at the surface and is defined as

$$Nu = \frac{hL}{k} \tag{13.23}$$

where h is the heat transfer coefficient [$W/(m^2 K)$], L is a characteristic length [m], k is the thermal conductivity [W/(m K)]. For flow through or normal to a pipe, the characteristic length is the pipe diameter (inner or outer, respectively).

Nu may be viewed as the ratio of convection to conduction for a layer of fluid. $Nu = 1$ means pure conduction. When $Nu > 1$, the heat transfer is enhanced by convection.

To determine the value of h for forced convection, follow these steps:

- choose from literature the appropriate Eq. (13.20);
- calculate the values of Re and Pr;

[f]Named after Ernst Kraft Wilhelm Nußelt (Nusselt in English), a German engineer (1882–1957).

- solve for Nu;
- determine h from Eq. (13.23).

Actual expressions of general equation (13.20) are determined from extensive experiments performed on particular surface geometries and types of flow. Such expressions are termed *empirical correlations* and are always accompanied by specifications regarding surface geometry and flow conditions [7, p. 410]. These specifications should not be neglected because they represent the limiting factors for the applicability of each expression.

As the heat exchange progresses, the temperature of the fluid changes which modifies the physical properties of the fluid (density, diffusivity, viscosity, etc.). Therefore it should be very clear for what temperature the properties have been considered in each empirical correlation. In general, all physical properties are evaluated either at the film temperature defined as

$$t_f = \frac{t_\infty + t_s}{2}$$

or at the fluid mean temperature

$$t_m = \frac{t_{\text{in}} + t_{\text{out}}}{2}$$

Typical correlations are presented in Table 13.2 [59]. For example, for turbulent flow along a flat plate, $Nu_L = 0.0365\, Re_L^{0.8} Pr^{1/3}$ for $Re_L > 5 \times 10^5$ and $0.6 \leq Pr \leq 60$. Here the properties of the fluid will be considered at t_f and L is the length of the plate in the direction of flow.

It is worth noting that (especially in turbulent flow) Nu does not depend on the shape of the tube's cross section. The same correlation can be used for any noncircular cross section using the concept of equivalent diameter d_e:

$$d_e = \frac{4A}{P}$$

where A is the inner area of the section and P is its wetted perimeter.

Table 13.2

Flow type	Surface type	Correlation	Conditions
External laminar flow	Plate	$Nu_L = 0.664\, Re_L^{1/2} Pr^{1/3}$	$Re_L < 5 \times 10^5$; L = length of plate in the direction of flow; t_f
	Cylinder (across)	$Nu_D = 0.25 + (0.4 Re_D^{1/2} + 0.06 Re_D^{2/3}) Pr^{0.37}$ (Sparrow)	$0.67 < Pr < 300$; $Re < 10^5$; D = cylinder diameter; t_f
External turbulent flow	Plate	$Nu_L = 0.0365\, Re_L^{0.8} Pr^{1/3}$	$Re_L > 5 \times 10^5$; $0.6 \leq Pr \leq 60$; t_f
Internal laminar flow	Tube	$Nu_d = 3.66 + \frac{0.0668(d/L) Re_d Pr}{1+0.04[(d/L) Re_d Pr]^{2/3}}$ (Hausen)	$Re_d < 2320$; d = inner diameter; L = tube length; t_m
Internal turbulent flow	Tube	$Nu_d = 0.024\, Re_d^{0.8} Pr^{0.4}$ for heating $Nu_d = 0.026\, Re_d^{0.8} Pr^{0.3}$ for cooling (Dittus-Boelter)	$Re_d > 2320$; $0.7 < Pr < 120$; d = inner diameter; t_m

If correlations for other configurations than the ones presented in Table 13.2 are required, reference should be made to specialized texts.

13.2.2. *The case of free convection*

It has been determined experimentally that, for the free (natural) convection, the following general correlation involving dimensionless numbers is applicable:

$$Nu = C(Gr\, Pr)^n \tag{13.24}$$

where Gr is the Grashof number and C and n are constants determined experimentally.

The **Grashof number**[g] is the dimensionless number associated with natural (free) convection. It is the ratio of buoyancy forces to

[g]Named after Franz Grashof (1826–1893), a German engineer, one of the founding directors of Verein Deutscher Ingenieure (VDI) in 1855.

the viscous forces, defined as

$$Gr = \frac{g\,\beta(t_s - t_\infty)L^3}{\nu^2} \tag{13.25}$$

where g is the acceleration due to gravity [m/s^2], β is the coefficient of volumetric expansion [m^3/(m^3 K)], L is a characteristic length [m], and ν is the kinematic viscosity [m^2/s].

In free convection the Grashof number plays the same role that is played by the Reynolds number in forced convection. For an ideal gas, $\beta = 1/T$ where T is the absolute temperature of the gas. For liquids and real gases β can be found in thermodynamic tables (e.g., [7, Tables HT-4 and HT-5]).

To determine the value of h for free convection, follow these steps:

- choose from literature the appropriate Eq. (13.24);
- calculate the values of Gr and Pr;
- solve for Nu;
- determine h from Eq. (13.23).

Some constants for use with Eq. (13.24) can be found in Table 13.3 [59]; physical properties are considered at the film temperature t_f.

Various functional correlations can also be found in literature.

The order of magnitude of convective heat transfer coefficients is presented in Table 13.4.

For free convection involving ambient air, a good average for h is 7 W/(m^2 K). A substantial increase in the value of h for liquids can be obtained by using *nanofluids* [60]. These are engineered colloidal suspensions of nanoparticles (metals, oxides, carbides, or carbon nanotubes) in a base fluid (water, ethylene glycol, oil). Nanofluids have properties that make them potentially useful in the field of heat transfer for applications including microelectronics and fuel cells.

Using the electrical analogy for convection, $\dot{Q} = h\,A_s(t_s - t_\infty)$ and

$$\dot{Q} = \frac{(t_s - t_\infty)}{\frac{1}{hA}} = \frac{(t_s - t_\infty)}{R_{t,\mathrm{conv}}} \tag{13.26}$$

Here

$$R_{t,\mathrm{conv}} = \frac{1}{hA} \tag{13.27}$$

Table 13.3

Surface type		**(*Gr Pr*)**	***C***	***n***
Vertical planes and cylinders	Vertical height $L < 1$ m	$<10^4$	0.36	1/5
		10^4–10^9	0.59	1/4
		$>10^9$	0.13	1/3
Horizontal cylinders	Diameter D used for L and $D < 0.2$ m	$>10^{-5}$	0.49	0
		10^{-5}–10^{-3}	0.71	1/25
		10^{-3}–1	1.09	1/10
		1–10^4	1.09	1/5
		10^4–10^9	0.53	1/4
		$>10^9$	0.13	1/3
Horizontal plates	Upper surface of heated plates or lower surface of cooled plates	10^5–2×10^7	0.54	1/4
		2×10^7–3×10^{10}	0.14	1/3
	Lower surface of heated plates or upper surface of cooled plates	10^5–10^{11}	0.58	1/5

Table 13.4

Convection type	**Description**	**Typical value of *h* W/(m² K)**
Natural (free)	Fluid motion induced by density differences	2–25 (gases) 50–1000 (liquids)
Forced	Fluid motion induced by pressure differences from a fan or pump	25–250 (gases) 100–20 000 (liquids)

is the **thermal resistance for convection** [7, p. 364]. Its measuring units are:

$$[R_{t,\text{cond}}]_{\text{SI}} = \text{K/W or } ^\circ\text{C/W};$$

$$[R_{t,\text{cond}}]_{\text{US Customary}} = (\text{h } ^\circ\text{F})/\text{Btu}$$

13.2.3. *The overall heat transfer*

The overall heat transfer considers the heat exchange between two flowing fluids separated by a wall. The one-dimensional heat transfer rate (in the x-direction) for the system in Fig. 13.7 may be expressed as

$$\dot{Q}_x = \frac{(t_{\infty,1} - t_{s,1})}{R_{t,\text{conv1}}} = \frac{(t_{s1} - t_{s2})}{R_{t,\text{cond}}} = \frac{(t_{s,2} - t_{\infty,2})}{R_{t,\text{conv2}}} = \frac{(t_{\infty,1} - t_{\infty,2})}{R_{\text{tot}}}$$

where $(t_{\infty,1} - t_{\infty,2})$ is the *overall temperature difference* and R_{tot} includes all thermal resistances (to conduction and convection).

$$R_{\text{tot}} = R_{t,\text{conv1}} + T_{t,\text{cond}} + R_{t,\text{conv2}} = \frac{1}{h_1 A} + \frac{L}{kA} + \frac{1}{h_2 A} \tag{13.28}$$

One can write

$$\dot{Q}_x = \frac{(t_{\infty,1} - t_{\infty,2})}{R_{\text{tot}}} = \frac{(t_{\infty,1} - t_{\infty,2})}{\frac{1}{h_1 A} + \frac{L}{k A} + \frac{1}{h_2 A}}$$

$$= \frac{A(t_{\infty,1} - t_{\infty,2})}{\frac{1}{h_1} + \frac{L}{k} + \frac{1}{h_2}} = UA(t_{\infty,1} - t_{\infty,2}) \tag{13.29}$$

Here U is the **overall heat transfer coefficient** [7, p. 366]. In general for a multi-layer plane wall

$$U = \frac{1}{\frac{1}{h_1} + \sum \left(\frac{L}{k}\right)_{\text{layer}} + \frac{1}{h_2}} \tag{13.30}$$

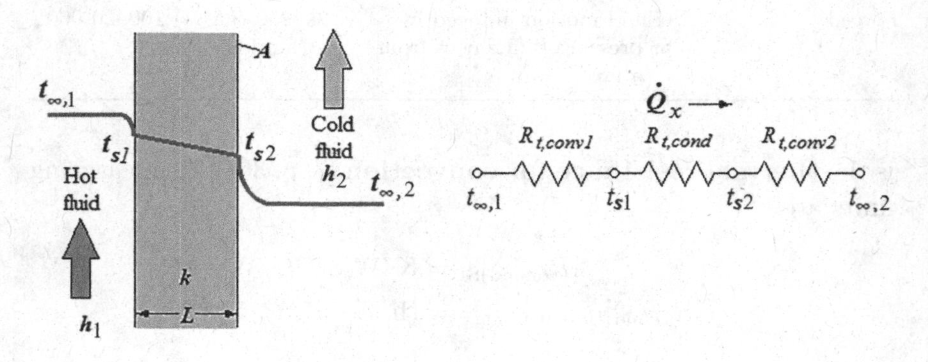

Fig. 13.7

or

$$\frac{1}{U} = \frac{1}{h_1} + \sum \left(\frac{L}{k} \right)_{\text{layer}} + \frac{1}{h_2} \tag{13.31}$$

For a cylindrical wall:

$$\dot{Q}_r = U'L(t_{\infty,1} - t_{\infty,2}) \tag{13.32}$$

where

$$\frac{1}{U'} = \frac{1}{2\pi r_1 h_1} + \sum \left\{ \frac{\ln(\frac{r_o}{r_i})}{2\pi\, k} \right\}_{\text{layer}} + \frac{1}{2\pi\, r_2 h_2} \tag{13.33}$$

In Eq. (13.31),

$$\frac{1}{U} = R \tag{13.34}$$

The values U are very important for the calculations of heat losses through a building element over a given area. The values of R from Eq. (13.34), also called the ***R*-values** — expressed in (°C m^2)/W or (h ft^2 °F)/Btu — are indicated in literature for some common constructions (such as house walls, double-glazed windows, etc.) and standardized thickness of each building element. The higher the R-value, the better the building insulation's effectiveness.

13.3. Radiation

Regardless of the temperature, all matter radiate heat. Radiation is transfer of heat through electromagnetic waves in the heat spectrum. Radiation does not require any direct contact or medium to propagate, because electromagnetic waves exist even in perfect vacuum. This is why we can receive the heat from the sun, so necessary for life on earth. Heat transfer by radiation occurs usually between solid surfaces, where electromagnetic waves are emitted by atomic vibration at a surface of a body. Liquids emit radiation by the random thermal motion of their component particles. Radiation from gases is also possible, but this phenomenon is important only at high temperatures. Solids radiate over a wide range of wavelengths, while some gases emit and absorb detectable radiation on certain wavelengths only. Bodies in thermal equilibrium emit and absorb energy in equal amounts to remain at a constant temperature.

For ordinary temperatures, most of the radiation is in the *infrared* region, well out of the visible spectrum. The infrared radiation passes through transparent substances like air and space without warming them to any significant extent. This is why thermal radiation from the sun warms the earth but not the space in between. Infrared radiation corresponds to wavelengths of 0.1 mm to 0.75 nm, or frequencies of (0.03 to 4) $\times 10^{14}$ Hz.

In physics, an ideal object that absorbs all electromagnetic radiation that falls onto it is called a **black body**. A black body is an idealized body which is a perfect absorber, and therefore also a perfect emitter. This concept is associated to the following properties:

- no radiation passes through it and none is reflected;
- it radiates every possible wavelength of energy, at a maximum rate possible for a given temperature;
- the amount and type of electromagnetic radiation emitted by a blackbody is isotropic[h] (diffuse emitter) and depends only on its absolute temperature.

At relatively low temperatures, a black body radiates mostly in the infrared (IR) spectrum. As the temperature of a black body increases, the maximum power density (W/m^2) of the radiation it emits approaches the visible spectrum (see Fig. 13.8). At about 3000 K the wavelengths become short enough to be visible: the body becomes "red hot", the maximum emissive power density occurs at 966 nm. At the temperature of the incandescent tungsten filament (about 2800 K), only about 10% of the emitted energy is in the visible range; this is the reason why incandescent lamps are so inefficient: the bulk of the radiation is in the infrared region (thermal radiation). At 4000 K, the peak emissive power density is at 724 nm (red). At higher temperatures the radiation extends more in the visible spectrum. At 6000 K — approximately the sun's surface temperature — the

[h]This means that radiation has the same intensity regardless of the direction of measurement.

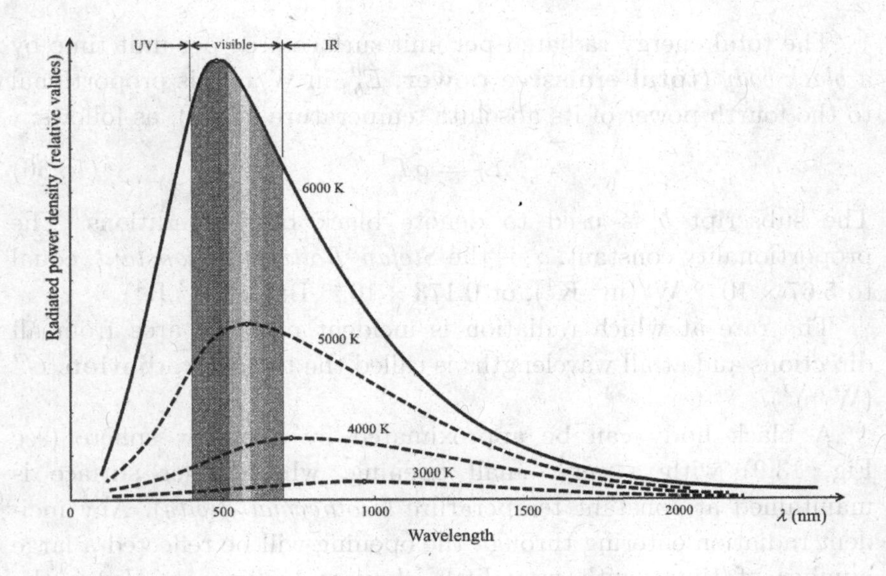

Fig. 13.8

radiated power reaches its peak in the visible spectrum (483 nm, blue), with about 43 % of the energy emitted in this range (and about 50 % in infrared). These values are based on *Wien's law*[i] (also called *Wien's displacement law* [7, p. 475]) stating that for a black body the wavelength of maximum emission is inversely proportional to its absolute temperature:

$$\lambda_{\text{peak}} T = 2.898 \times 10^{-3} \text{ m K} \tag{13.35}$$

This shift in the wavelength at which radiant power is a maximum is very important for the use of solar energy in a greenhouse. The glass allows the solar radiation in, but does not let the heat radiation out. This is possible because the two radiations are in very different wavelength ranges — say for 5700 K and 300 K — and ordinary glass is transparent to light but opaque to infrared (and UV) radiation. Greenhouses only work because λ_{peak} varies with temperature.

[i]Discovered by Wilhelm Carl Werner Otto Fritz Franz Wien (1864–1928), a German physicist.

The total energy radiated per unit surface area per unit time by a *black body* (**total emissive power**, $\dot{E}_b''$, in W/m^2) is proportional to the fourth power of its absolute temperature T_s (K), as follows:

$$\dot{E}_b'' = \sigma T_s^4 \tag{13.36}$$

The subscript b is used to denote black body conditions. The proportionality constant, σ, is the *Stefan–Boltzmann constant*, equal to 5.67×10^{-8} W/(m^2 K^4), or 0.173×10^{-8} Btu/(h ft^2 R^4).

The rate at which radiation is incident per unit area from all directions and at all wavelengths is called the **total irradiation**, $\dot{G}''$ (W/m^2).

A black body can be approximated by a hollow sphere (see Fig. 13.9) with a very small opening, whose inner surface is maintained at constant temperature (*isothermal cavity*). Any incident radiation entering through the opening will be reflected a large number of times with very little chances to emerge. Hence the incident radiation is practically entirely absorbed. At the same time, the cavity maintained at some finite temperature will radiate energy through the small hole according to the surface temperature. The radiation emitted by the interior surface is reflected many times and eventually fills the cavity uniformly and must be of the same form as the radiation emerging from the opening. Therefore any small surface placed in the cavity will receive radiation isotropically, for which

$$\dot{G}'' = \dot{E}_b''$$

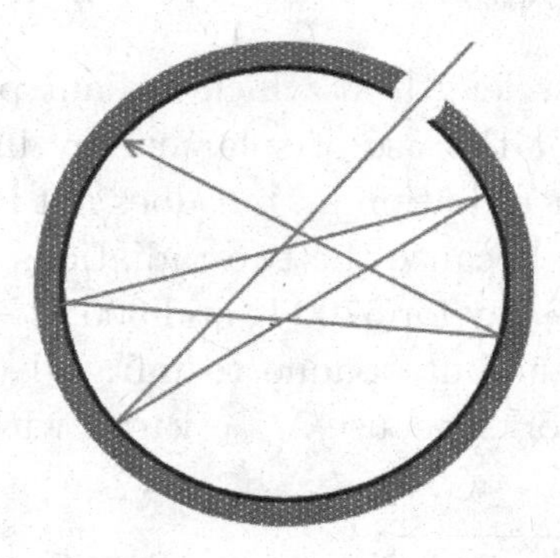

Fig. 13.9

It is worth noting that black body radiation conditions exist in the cavity regardless of whether the inner walls are reflective or absorbing [7, p. 474].

Real surfaces are not black and they emit less radiation than a black surface. Objects presenting such surfaces are called *gray bodies.* For such a body, the radiated heat per unit time ($\dot{E}$, in W) depends on the material properties represented by ε, the *emissivity* of the material. The rate of energy radiated is therefore:

$$\dot{E} = \varepsilon \sigma A T_s^4 \tag{13.37}$$

where σ is the Stefan–Boltzmann constant [W/(m^2 K^4)], A is the surface area of the object [m^2], T_s is the absolute temperature of the surface [K]. The emissivity ε of a surface is the ratio of the emissive power $\dot{E}''$ of that surface to the emissive power of a black body $\dot{E}_b''$ at the same temperature. The values of *emissivity* are between 0 and 1, depending on the surface material and finish. A black body — perfect emitter of heat energy — has an emissivity value of 1. A material with an emissivity value of 0 would be considered a perfect thermal mirror. Tables of total emissivity values can be found in literature (see also the Appendix E, Table A.9); they usually contain ranges of emissivity values for real materials because, for any given temperature, emissivity can be affected by surface roughness or finish.

In the real world, the thermal radiation striking a body can be *absorbed* by the body, *reflected* from the body, or *transmitted* through the body. The fraction of the incident radiation which is absorbed by the body is called *absorptivity* (α), the fraction reflected is called *reflectivity* (ρ) and the fraction transmitted is called *transmissivity* (τ). The sum of these fractions should be unity, i.e.

$$\alpha + \rho + \tau = 1 \tag{13.38}$$

Shiny materials typically reflect radiant heat, just as they reflect visible light. Dark materials typically absorb heat, just as they absorb visible light. In general,

$$\begin{aligned}\text{Incident energy} = \text{Absorbed energy} &+ \text{Reflected energy}\\ &+ \text{Transmitted energy}\end{aligned}$$

The energy absorbed by the object increases its thermal energy; the transmitted and reflected energy do not. In order for the temperature of the object to remain constant, the object must radiate the same amount of energy as it absorbs

$$\text{Emitted energy} = \text{Absorbed energy}$$

Therefore, objects that are good absorbers are good emitters and objects that are poor absorbers are poor emitters. Therefore

$$\text{Incident energy} = \text{Emitted energy} + \text{Reflected energy} + \text{Transmitted energy}$$

The heat transfer rate to a body will be equal to the difference between the energy absorbed and the energy emitted. The heat exchange between a gray body of a small surface A with emissivity ε and temperature T_s and a black, surrounding surface at temperature T_{surr}, ($T_{\text{surr}} > T_s$) separated by a gas, is

$$\dot{Q} = \varepsilon\sigma A(T_{\text{surr}}^4 - T_s^4) \tag{13.39}$$

Equation (13.39) holds even if the surrounding are not black but gray, provided the surroundings are large compared with the body.

The heat exchange between two finite gray bodies 1 and 2 depend on their radiative properties taken into account by the *emissivity factor* ε_{12}.

Also, the radiation exchange between two or more solids depends on surface geometries and orientations. The *view factor* F_{ij} is defined as the fraction of the radiation leaving surface i that is intercepted by surface j. It can be shown that for surfaces i and j, $A_i F_{ij} = A_j\ F_{ji}$.

The heat exchange between two finite gray bodies 1 and 2 at temperatures T_1 and T_2, respectively, ($T_1 > T_2$) is

$$\dot{Q} = \sigma\varepsilon_{12}F_{12}A(T_1^4 - T_2^4) \tag{13.40}$$

The factors ε_{12} and F_{12} depend on the surface arrangement of the two bodies.

The value of ε_{12} is calculated with appropriate formulas, depending on the mutual position of the two surfaces that exchange heat by radiation (see also [7, p. 498]). For example, for two parallel walls with emissivities ε_1 and ε_2,

$$\varepsilon_{12} = \frac{1}{\frac{1}{\varepsilon_1} + \frac{1}{\varepsilon_2} - 1}$$

For two long coaxial cylinders

$$\frac{1}{\varepsilon_{12}} = \frac{1}{\varepsilon_1} + \frac{d_1}{d_2}\left(\frac{1}{\varepsilon_2} - 1\right)$$

where ε_1 and ε_2 are the emissivities, d the diameter, and subscripts 1 and 2 refer to the inner and outer cylinders, respectively.

View factors can be found in literature for situations that occur frequently (see [7, pp. 490–491]).

13.4. Heat exchangers

A **heat exchanger** is a device built to allow heat transfer between two fluids that are at different temperatures. They are used for a variety of industrial processes requiring heat transfer in power plants (classic and nuclear), petrochemical plants, and natural-gas processing plants. They are an inherent part of HVAC installations used in dwellings, cars, airplanes, rockets and space stations. They are also integrated in power distribution systems and, on a smaller scale, they are used for cooling of electric or electronic components including computers. Heat exchangers work because heat naturally flows from higher temperatures to lower temperatures. There are three main types of heat exchangers:

- *recuperative* — where the fluids are separated by a solid wall so they do not mix. Heat is transferred from fluids to wall by convection, and through the wall itself by conduction;
- *regenerative* — where hot and cold fluids occupy the same space alternatively; the space contains a matrix of material with high thermal massivity that plays the role of a sink or a source for heat flow;
- *evaporative* (such as cooling towers) — where a liquid is cooled by evaporation in a gaseous coolant.

The recuperative type of heat exchangers is the most commonly used in practice. For a recuperative heat exchanger, assuming steady-state, steady-flow conditions, the heat transfer rate $\dot{Q}$ in (W) between the hot fluid (index h) and the cold fluid (index c) can be written as

$$\dot{Q} = \dot{m}_h c_{p,h}(t_{h,i} - t_{h,o}) \tag{13.41}$$

$$\dot{Q} = \dot{m}_c c_{p,c}(t_{c,o} - t_{c,i}) \tag{13.42}$$

where $\dot{m}$ is mass flow rate [kg/s], c_p is the specific heat [J/(kg K)], t_i and t_o are the inlet and, respectively, outlet temperatures, in [°C or K].

Introducing the *heat capacity rates* $\dot{C}_h$ and $\dot{C}_c$ [7, p. 448], in [W/K] or [W/°C], defined as

$$\dot{C}_h = \dot{m}_h \, c_{p,h} \tag{13.43}$$

$$C_c = \dot{m}_c \, c_{p,c} \tag{13.44}$$

Equations (13.30) and (13.31) can be written as

$$\dot{Q} = \dot{C}_h(t_{h,i} - t_{h,o}) \tag{13.45}$$

$$\dot{Q} = \dot{C}_c(t_{c,o} - t_{c,i}) \tag{13.46}$$

At the same time, for a recuperative heat exchanger, the convection rate equation can be written as

$$\dot{Q} = U A \Delta t_m \tag{13.47}$$

where U is the overall heat transfer coefficient [W/(m^2 °C)], A is the heat transfer area [m^2], Δt_m is the mean temperature difference, in [K] or [°C]. The product UA is called **overall conductance**. It is worth noting that the value of U is not a constant for a given type of heat exchanger; it depends upon many functional parameters: nature of fluids, temperature difference, degree of turbulence, etc. Because the temperatures of the two fluids change along the heat exchanger, the temperature difference Δt between the two fluids is not constant, so a mean value Δt_m has to be considered. This mean depends on the flow arrangement through the heat exchanger: *parallel-flow*, *counter-flow*, or *cross-flow*. Assuming that, for given operating conditions, U and $\dot{C}$ are constant along the entire flow length of the heat exchanger, the appropriate mean temperature difference required in Eq. (13.36) is the **logarithmic mean temperature difference** (or *log mean temperature difference*[j] or even *LMTD*) Δt_{lm}. A derivation of its expression for various flow arrangements can be found in [7, p. 450; 61, p. 108].

In the case of **parallel flow**, the fluids flow in the same direction through the heat exchanger (see Fig. 13.10).

The expression of Δt_{lm} is

$$\Delta t_{\text{lm}} = \frac{\Delta t_{\text{max}} - \Delta t_{\text{min}}}{\ln \frac{\Delta t_{\text{max}}}{\Delta t_{\text{min}}}} \tag{13.48}$$

where Δt_{lm} and Δt_{lm} are the **endpoint temperatures**. For parallel flow, these temperatures are

$$\Delta t_{\text{max}} = t_{h,i} - t_{c,i}, \quad \Delta t_{\text{min}} = t_{h,o} - t_{c,o} \tag{13.49}$$

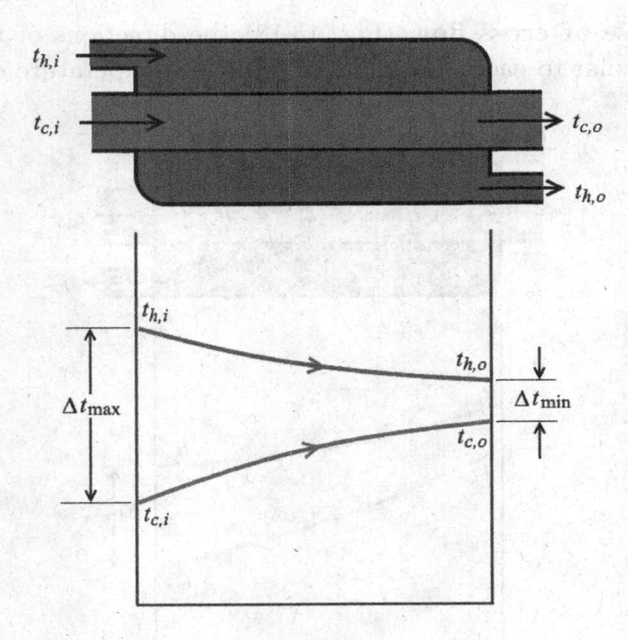

Fig. 13.10

[j]See also Appendix C.

Notice that for parallel flow the maximum temperature difference is always on the inlet side.

In the case of **counterflow**, the fluids flow in opposite directions through the heat exchanger. This is the most efficient pattern when comparing heat transfer rates per unit surface area. The mean temperature difference is calculated also with Eq. (13.48), where (see Fig. 13.11)

$$\Delta t_{\mathrm{max}} = \max\{(t_{h,i} - t_{c,o}), (t_{h,o} - t_{c,i})\},$$

$$\Delta t_{\mathrm{min}} = \min\{(t_{h,i} - t_{c,o}), (t_{h,o} - t_{c,i})\} \tag{13.50}$$

For this arrangement, it is possible that $t_{c,o} > t_{h,o}$, which cannot be attained with parallel flow. For the special case $(t_{h,i} - t_{c,o}) = (t_{h,o} - t_{c,i})$,

$$\Delta t_{\mathrm{lm}} = (t_{h,i} - t_{c,o}) = (t_{h,o} - t_{c,i})$$

In the case **of cross-flow** (Fig. 13.12) the directions of fluids are perpendicular to each other. The true mean temperature difference

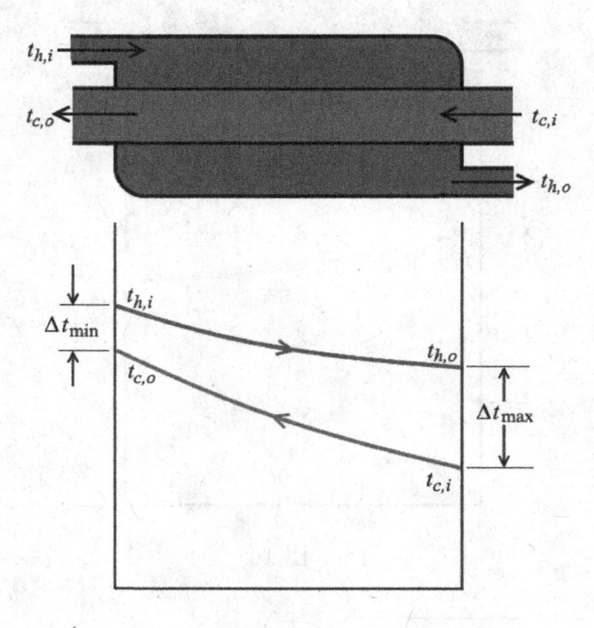

Fig. 13.11

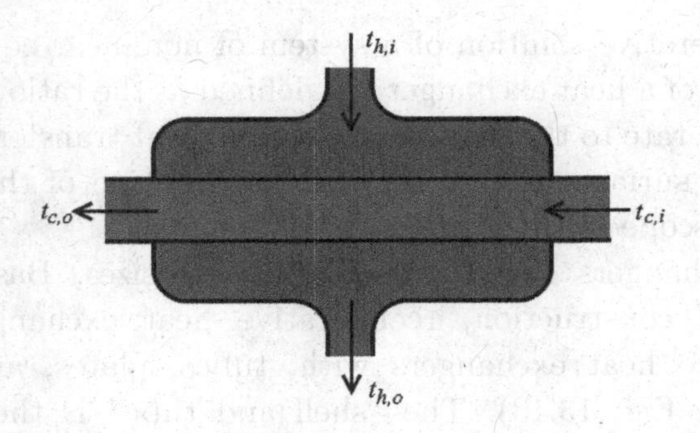

Fig. 13.12

is calculated in this case using the Δt_{lm} corrected by a *correction factor* F:

$$\Delta t_{\mathrm{lm}} = F\,\Delta t_{\mathrm{lm,counterflow}} \tag{13.51}$$

The values of F can be found in literature presented graphically for common heat exchanger configurations (e.g., [7, p. 454]). It can be seen from those charts that if the temperature change of one fluid is zero or at least negligible, then $F = 1$. In this case, the exchanger behavior is independent of the specific flow arrangement. This happens when one of the fluids undergoes a phase change or has a much higher heat capacity rate than the other one.

Notice that Eqs. (13.45)–(13.48) all require that both inlet and outlet temperatures are known. This is the disadvantage of the *log mean temperature difference* (**LMTD**) *method.* The LMTD method is easy to use if the required conductance needs to be determined (a design-type of problem). For situations where the inlet temperatures and U are known and the outlet temperatures are required (a simulation-type of problem), the LMTD method is inconvenient. In this case, a nonlinear system of two equations (e.g., (13.45) and (13.47)) need to be solved for two unknowns: $t_{h,o}$ and $t_{c,o}$. The solution requires application of an iterative method and convergence issues are possible.

A more convenient method for predicting the outlet temperatures is the *effectiveness-number of transfer units* (**ε-NTU**) *method.* This method can predict the outlet temperatures without resorting to a numerical iterative solution of a system of nonlinear equations. The effectiveness of a heat exchanger ε is defined as the ratio of the actual heat transfer rate to the maximum possible heat transfer rate if there were infinite surface area. A detailed presentation of this method is beyond the scope of this course.

Heat exchangers cover a wide range of sizes. Based on their geometry of construction, recuperative heat exchangers can be classified into heat exchangers with: tubes, plates, and extended surfaces (see Fig. 13.13). The "shell and tube" is the most common type of industrial heat ex-changer used in the petroleum, chemical and HVAC industries. It contains a number of parallel tubes inside a shell (Fig. 13.13(a) shows the shell-and-tube heat exchanger with one tube pass). A two-pass alternative is presented in Fig. 13.13(b).

For a given LMTD, the heat transfer rate can be increased by increasing the surface area for heat exchange. One way of achieving

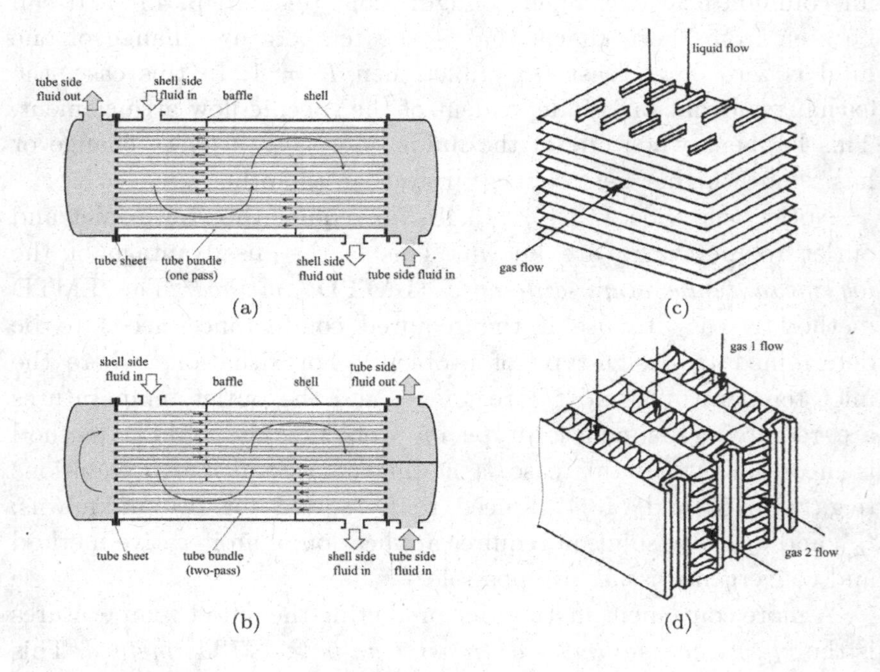

Fig. 13.13

this is through the use of **extended** or **finned surfaces**. Extended surface heat exchangers are designed and constructed to suit specific requirements with the combination of fin and tube materials and fin type selected to suit the application. In general, the extended surfaces are placed on the side of the fluid with the lowest convection coefficient; therefore, on liquid-to-gas heat exchangers, the finned surfaces will be on the gas side (Fig. 13.13(c)). However, in many applications — like car radiators — fins, ripples and dimples are placed also on the liquid side, to increase turbulence and to create a pressure-resistant structure (flat tubes). Plate-and-fin gas-to-gas heat exchangers (Fig. 13.13(d)) are also called *compact heat exchangers*; they may include well over 1000 m^2 of heat transfer surface per cubic meter of exchanger.

Heat transfer is a vast field of study. In this section we covered the very basics of the mechanisms of heat transfer. The interested student will be able to find in literature (e.g., [1, 7, 58, 62]), a plethora of information to discover more about this exciting engineering field.

Heat transfer — problems

13.1. The wall of an industrial furnace consists of three side-by-side layers of brick: at the interior, a layer of fire brick of thickness $L_{\text{fb}} = 23$ cm and thermal conductivity $k_{\text{fb}} = 1$ W/(m K); a second layer of insulating brick of thickness $L_{\text{ib}} = 15$ cm and conductivity $k_{\text{ib}} = 0.15$ W/(m K); at the exterior, a layer of regular building brick with a thickness $L_{\text{bb}} = 20$ cm and the conductivity $k_{\text{bb}} = 0.6$ W/(m K). The temperature of the inner surface of the furnace is 850 °C and the temperature of the external surface is maintained at 50 °C. Determine: (a) the heat dissipated through the composite wall and the intermediate temperatures; (b) the heat dissipated and the intermediate temperatures assuming the presence of a 7-mm thick layer of air with $k_{\text{air}} = 0.06$ W/(m K) between the fire brick and the insulating brick layers.

13.2. Determine the heat flux dissipated per unit length of an insulated steam pipe. The pipe has an internal diameter of 134.5 mm, a wall thickness of 3.4 mm and a thermal conductivity of 64 W/(m K). The insulation consists of two layers, one with the thickness $\delta_1 = 30$ mm and thermal conductivity $k_1 = 0.037$ W/(m K), the other with the thickness $\delta_2 = 50$ mm and thermal conductivity $k_2 = 0.14$ W/(m K). The temperature of the inner surface of the pipe is $t_i = 280$ °C and that of the outer surface of insulation is $t_o = 40$ °C. What happens to the heat flux if the layers of insulation are applied in reversed order?

13.3. A 3.5-m-long by 1-m-high vertical plate is maintained at a constant temperature of 140 °C. The back of the plate is insulated and the ambient air has a temperature of 20 °C. Determine the convective heat transfer coefficient and the heat transfer rate from the plate. Neglect radiation.

13.4. Water with an average temperature of $t_m = 90$ °C flows with the speed of 0.25 m/s through a pipe of length $L = 2.5$ m and inner diameter $d = 21$ mm. The average temperature of the wall is $t_s = 110$ °C. Determine the average heat transfer rate from the wall to the water.

13.5. A pipe is cooled by means of forced convection from a stream of water. The flow moves across the pipe at 1.2 m/s (upstream velocity) and its bulk temperature is 15 °C. The pipe is 2 m long, with a diameter of 6 mm and a temperature of 35 °C, determine the coefficient of heat transfer and the heat transfer rate.

13.6. Two parallel plates with emissivities $\varepsilon_1 = 0.75$ and $\varepsilon_2 = 0.65$ have temperatures $T_1 = 600$ K and $T_2 = 850$ K. Determine the radiant heat exchange between the two finite grey bodies and also calculate the radiant energy transfer if the plates were black.

Appendix A: The Derivation of the Adiabatic Condition

From the first law, for one unit mass of substance ($m = 1$):

$$\delta q - \delta w = du$$

When $q = 0$ (adiabatic process),

$$\begin{aligned} du &= -\delta w \\ du &= -p\,dv \end{aligned} \tag{A.1}$$

From the definition of specific heats, using (A.1), one obtains:

$$\begin{aligned} c_v &= \frac{du}{dT} = \frac{-p\,dv}{dT} \\ dT &= -\frac{p\,dv}{c_v} \end{aligned} \tag{A.2}$$

Also, from Mayer's relation,

$$R = c_p - c_v \tag{A.3}$$

For an ideal gas,

$$pv = RT \tag{A.4}$$

Differentiating (A.4) one obtains:

$$v\,dp + p\,dv = R\,dT$$

Replacing R and dT with their equivalents from (A.3) and (A.2), one obtains

$$v\,dp + p\,dv = (c_p - c_v)\left(-\frac{p}{c_v}dv\right)$$

Then

$$\frac{v\,dp + p\,dv}{c_p - c_v} = -\frac{p}{c_v}dv \quad \text{or} \quad \frac{v\,dp + p\,dv}{c_p - c_v} + \frac{p}{c_v}dv = 0$$

Dividing by pv, one obtains

$$\frac{1}{c_p - c_v}\frac{dp}{p} + \frac{1}{c_p - c_v}\frac{dv}{v} + \frac{1}{c_v}\frac{dv}{v} = 0$$

$$\frac{1}{c_p - c_v}\frac{dp}{p} + \left(\frac{1}{c_p - c_v} + \frac{1}{c_v}\right)\frac{dv}{v} = 0$$

$$\frac{1}{c_p - c_v}\frac{dp}{p} + \frac{c_p}{c_v}\frac{1}{c_p - c_v}\frac{dv}{v} = 0$$

Multiplying the equation by $(c_p - c_v)$ one obtains

$$\frac{dp}{p} + \frac{c_p}{c_v}\frac{dv}{v} = 0$$

For a perfect gas, the ratio $\frac{c_p}{c_v}$ is a constant which will be denoted by k. Hence

$$\frac{dp}{p} + k\frac{dv}{v} = 0$$

This is a separable differential equation. Integrating to find the solution, one obtains

$$\int \frac{dp}{p} + k\int \frac{dv}{v} = cst$$

$$\ln p + k \ln v = cst$$

$$\ln p + \ln v^k = cst \tag{A.5}$$

$$\ln(pv^k) = cst$$

$$pv^k = cst$$

Appendix B: Derivation of the Expression for the *Re* Number

The following convention will be adopted:

- [*quantity*] means "the dimension for that quantity"; e.g., [*mass*] = the dimension for mass.
- V = dimension velocity;
- t = dimension time;
- l = dimension length (linear dimension);
- ν = dimension viscosity (kinematic);
- μ = dimension viscosity (dynamic).

Let us remember first that the *Re* number represents the ratio of the inertia forces to the viscous forces.

- The inertia forces show how much a particular fluid resists any change in motion:

$$[\text{force}] = [\text{mass}] \times [\text{acceleration}]$$

$$[\text{mass}] = [\text{density}] \times [\text{volume}] = \rho l^3$$

$$[\text{acceleration}] = \frac{[\text{velocity}]}{[\text{time}]} = \frac{V}{t} = \frac{V}{\frac{l}{V}} = \frac{V^2}{l}$$

Therefore

$$[\text{inertia forces}] = \rho\, l^3 \frac{V^2}{l} = \rho l^2 V$$

- The viscous forces arise from the shear stress in the fluid:

$$[\text{viscous force}] = [\text{shear stress}] \times \begin{bmatrix}\text{surface area over which}\\ \text{the shear stress acts}\end{bmatrix}$$

$$[\text{shear stress}] = [\text{viscosity}] \times [\text{rate of share strain}]$$

$$[\text{rate of shear strain}] = [\text{velocity gradient}] = \frac{V}{l}$$

$$[\text{shear stress}] = \mu \frac{V}{l}$$

$$[\text{surface area}] = l^2$$

$$[\text{viscous force}] = \mu \frac{V}{l} l^2 = \mu V l$$

Therefore, the Reynolds number is

$$Re = \frac{[\text{inertia force}]}{[\text{viscous force}]} = \frac{\rho l^2 V}{\mu V l} = \frac{\rho l V}{\mu} = \frac{l V}{\frac{\mu}{\rho}} = \frac{l V}{\nu}$$

Appendix C: The Logarithmic Mean Concept

An interesting thing is that the concept of log mean difference — which we introduced in association with the temperature difference in the heat exchanger — was already used (without being declared as such) for the expression of the mean radius when calculating the heat rate transfer by conduction through a hollow tube. Therefore,

$$\dot{Q}_r = \frac{2\pi\, L\, k\, (t_i - t_o)}{\ln \frac{r_o}{r_i}} \tag{13.12}$$

Remember that for a planar wall,

$$\dot{Q}_x = \frac{k}{L} A(t_{\text{hot}} - t_{\text{cold}}) \tag{13.5}$$

By similarity, using a mean radius, Eq. (13.5) could be adapted for a cylindrical wall considering the material thickness $(r_o - r_i)$, and the cylindrical area $A = (2\pi r_{\text{mean}})L$. Then

$$\dot{Q}_r = \frac{(2\pi\, r_{\text{mean}})Lk(t_i - t_o)}{(r_o - r_i)}$$

Comparing the above equation with Eq. (13.12) it follows that:

$$\frac{r_{\text{mean}}}{(r_o - r_i)} = \frac{1}{\ln \frac{r_o}{r_i}}$$

or

$$r_{\text{mean}} = \frac{(r_o - r_i)}{\ln \frac{r_o}{r_i}}$$

which is an expression similar to that used for the log mean temperature difference; therefore we could call r_{mean} a "log mean radius".

The same concept of log mean difference can be extended also to pressure, to calculate a logarithmic mean pressure difference, useful for determining the friction coefficient for variable flow in parallel flow channels of the plate heat exchangers [63].

Appendix D: Nomenclature

A	area
c	velocity of sound
c_p	specific heat at constant pressure
c_v	specific heat at constant volume
COP	coefficient of performance
cst	constant
D	diameter
E	energy
F	force
g	gravitational acceleration
Gr	Grashof number
h	specific enthalpy; convection heat transfer coefficient
HHV	higher heating value
I	electric current
k	adiabatic exponent (specific heat ratio (c_p/c_v); thermal conductivity
LHV	lower heating value
m	mass
M	molar mass; Mach number
n	number of moles; polytropic exponent
Nu	Nusselt number
p	pressure
P	power
p_{cr}	critical pressure

Pr Prandtl number
q heat transfer per unit mass
Q total heat transfer
$\dot{Q}$ heat transfer rate
r compression ratio
R gas constant
$\bar{R}$ universal gas constant
r_c cutoff ratio
r_p pressure ratio
Re Reynolds number
s specific entropy
S total entropy
t empirical temperature
T thermodynamic (absolute) temperature
$T_{\rm cr}$ critical temperature
u specific internal energy
U total internal energy
v specific volume
V total volume; voltage
V velocity (speed) of flow
w work per unit mass
W total work
$\dot{W}$ power
x quality of saturated mixture
z elevation

Greek letters
α absorptivity
β coefficient of volumetric expansion
δ infinitesimal change in path quantity
Δ finite change in quantity
ε emissivity
η efficiency
μ dynamic viscosity
ν kinematic viscosity
ρ density

σ	Stefan–Boltzmann constant
τ	time
φ	relative humidity
ω	specific or absolute humidity

Subscripts

a	air
C	cold (low temperature)
f	saturated liquid
fg	difference in property between saturated liquid and saturated vapor
g	saturated vapor
H	hot (high temperature)
in	inlet conditions
out	outlet conditions
s	isentropic
th	thermal
v	water vapor
1	initial or inlet state
2	final or outlet state

Superscript

$\dot{}$ (overdot)	quantity per unit time

Appendix E: Property Tables and Charts

Table A.1. Ideal-gas specific heats of various common gases at 300 K.

Gas	Formula	Molar mass M [kg/kmol]	Gas constant R [J/(kg K)]	c_p [kJ/(kg K)]	c_v [kJ/(kg K)]	k
Air	—	28.97	287.0	1.004	0.717	1.400
Butane	C_4H_{10}	58.124	143.0	1.718	1.575	1.091
Carbon dioxide	CO_2	44.01	188.9	0.846	0.658	1.287
Carbon monoxide	CO	28.011	296.8	1.043	0.746	1.398
Ethane	C_2H_6	30.07	276.5	1.767	1.491	1.185
Helium	He	4.003	2077.0	5.193	3.116	1.667
Hydrogen	H_2	2.016	4124.2	14.321	10.197	1.404
Methane	CH_4	16.043	518.3	2.250	1.732	1.299
Nitrogen	N_2	28.013	296.8	1.038	0.741	1.400
Octane	C_8H_{18}	114.228	72.8	1.711	1.638	1.044
Oxygen	O_2	31.999	259.8	0.920	0.660	1.394
Propane	C_3H_8	44.097	188.5	1.681	1.492	1.126
Steam	H_2O	18.015	461.5	1.869	1.407	1.328

Source: The gas constant is calculated from $R = \bar{R}/M$ where $\bar{R} =$ 8314.41 J/kmol. Specific heats c_p have been calculated using the polynomial expression $c_p = a + bT + cT^2 + dT^3$ where the values of the coefficients a, b, c, d were taken from http://www.wiley.com/college/moran/CL_0471465704_S/user/tables/TABLE3S/table3sframe.html. (Accessed 17 March 2016). Errors are less than 1 %. $c_v = c_p - R; k = c_p/c_v$.

Table A.2. Saturated water — temperature table.

		Specific volume [m³/kg]		Internal energy [kJ/kg]		Enthalpy [kJ/kg]		Entropy [kJ/(kg K)]	
t [°C]	p [bar]	v_f	v_g	u_f	u_g	h_f	h_g	s_f	s_g
0.01	0.0061	0.001	205.99	0	2374.9	0.00	2500.9	0	9.1555
5	0.0087	0.001	147.01	21.02	2381.8	21.02	2510.1	0.0763	9.0248
10	0.0123	0.001	106.3	42.02	2388.6	42.02	2519.2	0.1511	8.8998
15	0.0171	0.001	77.875	62.98	2395.5	63.00	2528.3	0.2245	8.7803
20	0.0234	0.001	57.757	83.91	2402.3	83.91	2537.4	0.2965	8.666
25	0.0317	0.001	43.337	104.83	2409.1	104.8	2546.5	0.3672	8.5566
30	0.0425	0.001	32.878	125.73	2415.9	125.7	2555.5	0.4368	8.452
35	0.0563	0.00101	25.205	146.63	2422.7	146.6	2564.5	0.5051	8.3517
40	0.0739	0.00101	19.515	167.53	2429.4	167.5	2573.5	0.5724	8.2555
45	0.096	0.00101	15.252	188.43	2436.1	188.4	2582.4	0.6386	8.1633
50	0.1235	0.00101	12.027	209.33	2442.7	209.3	2591.3	0.7038	8.0748
55	0.1576	0.00102	9.5643	230.24	2449.3	230.3	2600.1	0.768	7.9898
60	0.1995	0.00102	7.6672	251.16	2455.9	251.2	2608.8	0.8313	7.9081
65	0.2504	0.00102	6.1935	272.09	2462.4	272.1	2617.5	0.8937	7.8296
70	0.0312	0.00102	5.0395	293.03	2468.9	293.2	2626.1	0.9551	7.754
75	0.386	0.00103	4.1289	313.99	2475.2	314.0	2634.6	1.0158	7.6812
80	0.4741	0.00103	3.4052	334.96	2481.6	335.0	2643.0	1.0756	7.6111
85	0.5787	0.00103	2.8258	355.95	2487.8	356.0	2651.3	1.1346	7.5434
90	0.7018	0.00104	2.3591	376.97	2494	377.0	2659.5	1.1929	7.4781
95	0.8461	0.00104	1.9806	398.0	2500	398.1	2667.6	1.2504	7.4151
100	1.0142	0.00104	1.6718	419.06	2506	419.2	2675.6	1.3072	7.3541
110	1.4338	0.00105	1.2093	461.26	2517.7	461.4	2691.1	1.4188	7.2381
120	1.9867	0.00106	0.8912	503.6	2528.9	503.8	2705.9	1.5279	7.1291
130	2.7028	0.00107	0.668	546.09	2539.5	546.4	2720.1	1.6346	7.0264
140	3.6154	0.00108	0.50845	588.77	2549.6	589.2	2733.4	1.7392	6.9293
150	4.7616	0.00109	0.39245	631.66	2559.1	632.2	2745.9	1.8418	6.8371
160	6.182	0.0011	0.30678	674.79	2567.8	675.5	2757.4	1.9426	6.7491
170	7.922	0.00111	0.24259	718.2	2575.7	719.1	2767.9	2.0417	6.665
180	10.028	0.00113	0.19384	761.92	2582.8	763.1	2777.2	2.1392	6.584
190	12.552	0.00114	0.15636	806.0	2589.0	807.4	2785.3	2.2355	6.5059
200	15.549	0.00116	0.12721	850.47	2594.2	852.3	2792	2.3305	6.4302
210	19.077	0.00117	0.10429	895.39	2598.3	897.6	2797.3	2.4245	6.3563
220	23.196	0.00119	0.08609	940.82	2601.2	943.6	2800.9	2.5177	6.284
230	27.971	0.00121	0.0715	986.81	2602.9	990.2	2802.9	2.6101	6.2128
240	33.469	0.00123	0.05971	1033.4	2603.1	1037.6	2803.0	2.702	6.1423
250	39.762	0.00125	0.05008	1080.8	2601.8	1085.8	2800.9	2.7935	6.0721
260	46.923	0.00128	0.04217	1129.0	2598.7	1135	2796.6	2.8849	6.0016

(*Continued*)

Table A.2. (*Continued*)

		Specific volume [m^3/kg]		Internal energy [kJ/kg]		Enthalpy [kJ/kg]		Entropy [kJ/(kg K)]	
t [°C]	p [bar]	v_f	v_g	u_f	u_g	h_f	h_g	s_f	s_g
270	55.03	0.0013	0.03562	1178.1	2593.7	1185.3	2789.7	2.9765	5.9304
280	64.166	0.00133	0.03015	1228.3	2586.4	1236.9	2779.9	3.0685	5.8579
290	74.418	0.00137	0.02556	1279.9	2576.5	1290	2766.7	3.1612	5.7834
300	85.879	0.0014	0.02166	1332.9	2563.6	1345	2749.6	3.2552	5.7059
310	98.651	0.00145	0.01834	1387.9	2547.1	1402.2	2727.9	3.351	5.6244
320	112.84	0.0015	0.01547	1445.3	2526.0	1462.2	2700.6	3.4494	5.5372
330	128.58	0.00156	0.01298	1505.8	2499.2	1525.9	2666.0	3.5518	5.4422
340	146.01	0.00164	0.01078	1570.6	2464.4	1594.5	2621.8	3.6601	5.3356
350	165.29	0.00174	0.0088	1642.1	2418.1	1670.9	2563.6	3.7784	5.211
360	186.66	0.0019	0.00695	1726.3	2351.8	1761.7	2481.5	3.9167	5.0536
370	210.44	0.00222	0.00495	1844.1	2230.3	1890.7	2334.5	4.1112	4.8012
373.95	220.64	0.00311	0.00311	2015.7	2015.7	2084.3	2084.3	4.407	4.407

Source: Adapted from "NISTIR 5078 — Thermodynamic Properties of Water: Tabulation from the IAPWS Formulation 1995 for the Thermodynamic Properties of Ordinary Water Substance for General and Scientific Use" available at https://nvlpubs.nist.gov/nistpubs/Legacy/IR/nistir5078.pdf (Accessed 31 July 2018). Internal energy was calculated as $u = h - pv$.

Note: Water, in pure or nearly pure form, is used in many industrial processes. Because of its importance, international standards exist for its thermophysical properties. These standards are set by the International Association for the Properties of Water and Steam (IAPWS; see www.iapws.org). Reliable data that meet the needs of industry are provided by the National Institute of Standards and Technology (NIST; see http://www.nist.gov/srd/nist10.cfm), based on software that implements IAPWS property standards (See also http://www.nist.gov/srd/upload/NISTIR5078.htm).

Table A.3. Saturated water — pressure table.

		Specific volume $[m^3/kg]$		Internal energy [kJ/kg]		Enthalpy [kJ/kg]		Entropy [kJ/(kg K)]	
p [bar]	t [°C]	vf	vg	uf	ug	hf	hg	sf	sg
0.04	28.96	0.001	34.79	121.4	2414.5	121.4	2553.7	0.4224	8.4734
0.06	36.16	0.00101	23.73	151.5	2424.2	151.5	2566.6	0.5208	8.329
0.08	41.51	0.00101	18.1	173.8	2431.4	173.8	2576.2	0.5925	8.2273
0.1	45.81	0.00101	14.67	191.8	2437.2	191.8	2583.9	0.6492	8.1488
0.12	49.42	0.00101	12.36	206.9	2442	206.9	2590.3	0.6963	8.0849
0.14	52.55	0.00101	10.69	220	2446.1	220	2595.8	0.7366	8.0311
0.16	55.31	0.00102	9.431	231.6	2449.8	231.6	2600.6	0.772	7.9846
0.18	57.8	0.00102	8.443	242	2453	242	2605	0.8036	7.9437
0.2	60.06	0.00102	7.648	251.4	2456	251.4	2608.9	0.832	7.9072
0.3	69.1	0.00102	5.228	289.2	2467.7	289.3	2624.5	0.9441	7.7675
0.4	75.86	0.00103	3.993	317.6	2476.3	317.6	2636.1	1.0261	7.669
0.6	85.93	0.00103	2.732	360	2489	359.9	2652.9	1.1454	7.5311
0.8	93.49	0.00104	2.087	391.6	2498.2	391.7	2665.2	1.233	7.4339
1	99.61	0.00104	1.694	417.4	2505.6	417.5	2674.9	1.3028	7.3588
1.2	104.78	0.00105	1.428	439.2	2511.7	439.4	2683.1	1.3609	7.2977
1.4	109.29	0.00105	1.2366	458.3	2516.9	458.4	2690	1.411	7.2461
1.6	113.3	0.00105	1.0914	475.2	2521.4	475.4	2696	1.4551	7.2014
1.8	116.91	0.00106	0.9775	490.5	2525.5	490.7	2701.4	1.4945	7.1621
2.0	120.21	0.00106	0.8857	504.5	2529.1	504.7	2706.2	1.5302	7.1269
3.0	133.52	0.00107	0.6058	561.1	2543.2	561.4	2724.9	1.6717	6.9916
4.0	143.61	0.00108	0.4624	604.2	2553.1	604.7	2738.1	1.7765	6.8955
6.0	158.83	0.0011	0.3156	669.7	2566.8	670.4	2756.1	1.9308	6.7592
8.0	170.41	0.00112	0.2403	720	2576	720.9	2768.3	2.0457	6.6616
10.0	179.88	0.00113	0.1944	761.4	2582.7	762.5	2777.1	2.1381	6.585
12.0	187.96	0.00114	0.1633	797	2587.8	798.3	2783.7	2.2159	6.5217
14.0	195.04	0.00115	0.1408	828.4	2591.8	830	2788.8	2.2835	6.4675
16.0	201.37	0.00116	0.1237	856.6	2594.8	858.5	2792.8	2.3435	6.4199
18.0	207.11	0.00117	0.1104	882.4	2597.2	884.5	2795.9	2.3975	6.3775
20.0	212.38	0.00118	0.0996	906.1	2599.1	908.5	2798.3	2.4468	6.339
30.0	233.85	0.00122	0.0667	1004.7	2603.2	1008.3	2803.2	2.6455	6.1856
40.0	250.35	0.00125	0.0498	1082.5	2601.7	1087.5	2800.8	2.7968	6.0696
50.0	275.58	0.00132	0.0325	1206	2589.9	1213.9	2784.6	3.0278	5.8901
60.0	295.01	0.00139	0.0235	1306.2	2570.5	1317.3	2758.7	3.2081	5.745
100	311	0.00145	0.018	1393.5	2545.2	1408.1	2725.5	3.3606	5.616
120	324.68	0.00153	0.0143	1473.1	2514.3	1491.5	2685.4	3.4967	5.4939
140	336.67	0.00161	0.0115	1548.4	2477.1	1571	2637.9	3.6232	5.3727
160	347.35	0.00171	0.0093	1622.3	2431.8	1649.7	2580.8	3.7457	5.2463
180	356.99	0.00184	0.0075	1699	2374.8	1732.1	2509.8	3.8718	5.1061
200	365.75	0.00204	0.0059	1786.4	2295	1827.2	2412.3	4.0156	4.9314
220.64	373.95	0.00311	0.00311	2015.7	2015.7	2084.3	2084.3	4.407	4.407

Source: Adapted from "NISTIR 5078 — Thermodynamic Properties of Water: Tabulation from the IAPWS Formulation 1995 for the Thermodynamic Properties of Ordinary Water Substance for General and Scientific Use" available at https://nvlpubs.nist.gov/nistpubs/Legacy/IR/nistir5078.pdf (Accessed 31 July 2018). Internal energy was calculated as $u = h - pv$. See also the note for Table A.2.

Table A.4. Superheated steam.

	$p = 0.10$ bar = 0.01 MPa				**$p = 0.50$ bar = 0.05 MPa**			
	$t_{sat} = 45.8$ °C				$t_{sat} = 81.3$ °C			
t [°C]	v [m^3/kg]	u [kJ/kg]	h [kJ/kg]	s [kJ/(kg K)]	v [m^3/kg]	u [kJ/kg]	h [kJ/kg]	s [kJ/(kg K)]
50	14.87	2443.3	2592.0	8.174	—	—	—	—
100	17.20	2515.5	2687.5	8.449	3.419	2511.5	2682.4	7.695
150	19.51	2587.9	2783.0	8.689	3.890	2585.7	2780.2	7.941
200	21.83	2661.3	2879.6	8.905	4.356	2660.0	2877.8	8.159
250	24.14	2736.1	2977.4	9.102	4.821	2735.1	2976.1	8.357
300	26.45	2812.3	3076.7	9.283	5.284	2811.6	3075.8	8.539
350	28.76	2890.0	3177.5	9.451	5.747	2889.4	3176.8	8.708
400	31.06	2969.3	3279.9	9.609	6.209	2968.9	3279.3	8.866
450	33.37	3050.3	3384.0	9.758	6.672	3049.9	3383.5	9.015
500	35.68	3132.9	3489.7	9.900	7.134	3132.6	3489.3	9.157
600	40.30	3303.3	3706.3	10.163	8.058	3303.1	3706.0	9.420
700	44.91	3480.8	3929.9	10.406	8.981	3480.6	3929.7	9.663

	$p = 1.0$ bar = 0.1 MPa				**$p = 2.0$ bar = 0.20 MPa**			
	$t_{sat} = 99.6$ °C				$t_{sat} = 120.2$ °C			
t [°C]	v [m^3/kg]	u [kJ/kg]	h [kJ/kg]	s [kJ/(kg K)]	v [m^3/kg]	u [kJ/kg]	h [kJ/kg]	s [kJ/(kg K)]
100	1.696	2506.2	2675.8	7.361	—	—	—	—
150	1.937	2582.9	2776.6	7.615	0.9599	2577.1	2769.1	7.281
200	2.172	2658.2	2875.5	7.836	1.0805	2654.6	2870.7	7.508
250	2.406	2733.9	2974.5	8.035	1.1989	2731.4	2971.2	7.710
300	2.639	2810.6	3074.5	8.217	1.3162	2808.8	3072.1	7.894
350	2.871	2888.7	3175.8	8.387	1.4330	2887.3	3173.9	8.064
400	3.103	2968.3	3278.6	8.545	1.5493	2967.1	3277.0	8.224
450	3.334	3049.4	3382.8	8.695	1.6655	3048.5	3381.6	8.373
500	3.566	3132.2	3488.7	8.836	1.7814	3131.4	3487.7	8.515
600	4.028	3302.8	3705.6	9.100	2.0130	3302.2	3704.8	8.779
700	4.490	3480.4	3929.4	9.342	2.2443	3479.9	3928.8	9.022
800	4.952	3665.0	4160.2	9.568	2.4755	3664.7	4159.8	9.248

	$p = 3.0$ bar = 0.30 MPa				**$p = 4.0$ bar = 0.40 MPa**			
	$t_{sat} = 133.5$ °C				$t_{sat} = 143.6$ °C			
t [°C]	v [m^3/kg]	u [kJ/kg]	h [kJ/kg]	s [kJ/(kg K)]	v [m^3/kg]	u [kJ/kg]	h [kJ/kg]	s [kJ/(kg K)]
150	0.6340	2571.0	2761.2	7.079	0.4709	2564.4	2752.8	6.931
200	0.7164	2651.0	2865.9	7.313	0.5343	2647.2	2860.9	7.172
250	0.7964	2728.9	2967.9	7.518	0.5952	2726.4	2964.5	7.380
300	0.8753	2807.0	3069.6	7.704	0.6549	2805.1	3067.1	7.568

(Continued)

Table A.4. (*Continued*)

	$p = 3.0$ bar $= 0.30$ MPa				$p = 4.0$ bar $= 0.40$ MPa			
	$t_{sat} = 133.5$ °C				$t_{sat} = 143.6$ °C			
t [°C]	v [m^3/kg]	u [kJ/kg]	h [kJ/kg]	s [kJ/(kg K)]	v [m^3/kg]	u [kJ/kg]	h [kJ/kg]	s [kJ/(kg K)]
350	0.9536	2885.9	3172.0	7.875	0.7140	2884.4	3170.0	7.740
400	1.0315	2966.0	3275.5	8.035	0.7726	2964.9	3273.9	7.900
450	1.1092	3047.5	3380.3	8.185	0.8311	3046.6	3379.0	8.051
500	1.1867	3130.6	3486.6	8.327	0.8894	3129.8	3485.5	8.193
600	1.3414	3301.6	3704.0	8.591	1.0056	3301.0	3703.2	8.458
700	1.4958	3479.5	3928.2	8.834	1.1215	3479.0	3927.6	8.701
800	1.6500	3664.3	4159.3	9.060	1.2373	3663.9	4158.8	8.927

	$p = 5.0$ bar $= 0.50$ MPa				$p = 8.0$ bar $= 0.80$ MPa			
	$t_{sat} = 151.8$ °C				$t_{sat} = 170.4$ °C			
t [°C]	v [m^3/kg]	u [kJ/kg]	h [kJ/kg]	s [kJ/(kg K)]	v [m^3/kg]	u [kJ/kg]	h [kJ/kg]	s [kJ/(kg K)]
200	0.4250	2643.3	2855.8	7.061	0.2609	2631.0	2839.7	6.818
250	0.4744	2723.8	2961.0	7.272	0.2932	2715.9	2950.4	7.040
300	0.5226	2803.2	3064.6	7.461	0.3242	2797.5	3056.9	7.235
350	0.5702	2883.0	3168.1	7.635	0.3544	2878.6	3162.2	7.411
400	0.6173	2963.7	3272.3	7.796	0.3843	2960.2	3267.6	7.573
450	0.6642	3045.6	3377.7	7.947	0.4139	3042.8	3373.9	7.726
500	0.7109	3129.0	3484.5	8.089	0.4433	3126.6	3481.3	7.869
600	0.8041	3300.4	3702.5	8.354	0.5019	3298.7	3700.1	8.135
700	0.8970	3478.5	3927.0	8.598	0.5601	3477.2	3925.3	8.379
800	0.9897	3663.6	4158.4	8.824	0.6182	3662.4	4157.0	8.606

	$p = 10.0$ bar $= 1.0$ MPa				$p = 14.0$ bar $= 1.4$ MPa			
	$t_{sat} = 179.9$ °C				$t_{sat} = 195.0$ °C			
t [°C]	v [m^3/kg]	u [kJ/kg]	h [kJ/kg]	s [kJ/(kg K)]	v [m^3/kg]	u [kJ/kg]	h [kJ/kg]	s [kJ/(kg K)]
200	0.2060	2622.2	2828.3	6.696	0.1430	2602.7	2803.0	6.498
250	0.2328	2710.4	2943.1	6.927	0.1636	2698.9	2927.9	6.749
300	0.2580	2793.6	3051.6	7.125	0.1823	2785.7	3040.9	6.955
350	0.2825	2875.7	3158.2	7.303	0.2003	2869.7	3150.1	7.138
400	0.3066	2957.9	3264.5	7.467	0.2178	2953.1	3258.1	7.305
450	0.3305	3040.9	3371.3	7.620	0.2351	3037.0	3366.1	7.459
500	0.3541	3125.0	3479.1	7.764	0.2522	3121.8	3474.8	7.605
600	0.4011	3297.5	3698.6	8.031	0.2860	3295.1	3695.4	7.873
700	0.4478	3476.2	3924.1	8.276	0.3195	3474.4	3921.7	8.118
800	0.4944	3661.7	4156.1	8.502	0.3529	3660.2	4154.3	8.346

(*Continued*)

Table A.4. (*Continued*)

	p = 20.0 bar = 2.0 MPa				**p = 3.0 bar = 3.0 MPa**			
	t_{sat} = 212.4 °C				t_{sat} = 233.90 °C			
t [°C]	v [m^3/kg]	u [kJ/kg]	h [kJ/kg]	s [kJ/(kg K)]	v [m^3/kg]	u [kJ/kg]	h [kJ/kg]	s [kJ/(kg K)]
225	0.1038	2628.5	2836.1	6.416	—	—	—	—
250	0.1115	2680.2	2903.2	6.548	0.0706	2644.7	2856.5	6.289
300	0.1255	2773.2	3024.2	6.768	0.0812	2750.8	2994.3	6.541
350	0.1386	2860.5	3137.7	6.958	0.0906	2844.4	3116.1	6.745
400	0.1512	2945.9	3248.3	7.129	0.0994	2933.5	3231.7	6.923
450	0.1635	3031.1	3358.2	7.287	0.1079	3021.2	3344.8	7.086
500	0.1757	3116.9	3468.2	7.434	0.1162	3108.6	3457.2	7.236
600	0.1996	3291.5	3690.7	7.704	0.1325	3285.5	3682.8	7.510
700	0.2233	3471.6	3918.2	7.951	0.1484	3467.0	3912.2	7.759
800	0.2467	3658.0	4151.5	8.179	0.1642	3654.3	4146.9	7.989

	p = 40.0 bar = 4.0 MPa				**p = 60.0 bar = 6.0 MPa**			
	t_{sat} = 250.4 °C				t_{sat} = 275.6 °C			
t [°C]	v [m^3/kg]	u [kJ/kg]	h [kJ/kg]	s [kJ/(kg K)]	v [m^3/kg]	u [kJ/kg]	h [kJ/kg]	s [kJ/(kg K)]
275	0.0546	2668.9	2887.3	6.231	—	—	—	—
300	0.0589	2726.2	2961.7	6.364	0.0362	2668.4	2885.5	6.070
350	0.0665	2827.4	3093.3	6.584	0.0423	2790.4	3043.9	6.336
400	0.0734	2920.7	3214.5	6.771	0.0474	2893.7	3178.2	6.543
450	0.0800	3011.0	3331.2	6.939	0.0522	2989.9	3302.9	6.722
500	0.0864	3100.3	3446.0	7.092	0.0567	3083.1	3423.1	6.883
600	0.0989	3279.4	3674.9	7.371	0.0653	3267.2	3658.7	7.169
700	0.1110	3462.4	3906.3	7.621	0.0736	3453	3894.3	7.425
800	0.1229	3650.6	4142.3	7.852	0.0817	3643.2	4133.1	7.658

	p = 80.0 bar = 8.0 MPa				**p = 100.0 bar = 10.0 MPa**			
	t_{sat} = 295.0 °C				t_{sat} = 311.0 °C			
t [°C]	v [m^3/kg]	u [kJ/kg]	h [kJ/kg]	s [kJ/(kg K)]	v [m^3/kg]	u [kJ/kg]	h [kJ/kg]	s [kJ/(kg K)]
300	0.0243	2592.3	2786.5	5.794	—	—	—	—
350	0.0300	2748.3	2988.1	6.132	0.0224	2699.6	2924.0	5.946
400	0.0343	2864.6	3139.4	6.366	0.0264	2833.1	3097.4	6.214
450	0.0382	2967.8	3273.3	6.558	0.0298	2944.5	3242.3	6.422
500	0.0418	3065.4	3399.5	6.727	0.0328	3047.0	3375.1	6.600
600	0.0485	3254.7	3642.4	7.022	0.0384	3242.0	3625.8	6.905
700	0.0548	3443.6	3882.2	7.282	0.0436	3434.0	3870.0	7.169
800	0.0610	3635.7	4123.8	7.518	0.0486	3628.2	4114.5	7.409

(*Continued*)

Table A.4. (*Continued*)

	$p = 125.0$ bar $= 12.5$ MPa				$p = 150.0$ bar $= 15.0$ MPa			
	$t_{sat} = 327.8$ °C				$t_{sat} = 342.16$ °C			
t [°C]	v [m^3/kg]	u [kJ/kg]	h [kJ/kg]	s [kJ/(kg K)]	v [m^3/kg]	u [kJ/kg]	h [kJ/kg]	s [kJ/(kg K)]
350	0.0161	2624.8	2826.6	5.713	0.0115	2520.8	2693.1	5.444
400	0.0200	2789.6	3040.0	6.043	0.0157	2740.6	2975.7	5.882
450	0.0230	2913.7	3201.4	6.275	0.0185	2880.7	3157.9	6.143
500	0.0256	3023.2	3343.6	6.465	0.0208	2998.4	3310.8	6.348
600	0.0303	3225.8	3604.6	6.783	0.0249	3209.3	3583.1	6.680
700	0.0346	3422.0	3854.6	7.054	0.0286	3409.8	3839.1	6.957
800	0.0387	3618.7	4102.8	7.297	0.0321	3609.2	4091.1	7.204

	$p = 175.0$ bar $= 17.5$ MPa				$p = 200.0$ bar $= 20.0$ MPa			
	$t_{sat} = 354.7$ °C				$t_{sat} = 365.8$ °C			
t [°C]	v [m^3/kg]	u [kJ/kg]	h [kJ/kg]	s [kJ/(kg K)]	v [m^3/kg]	u [kJ/kg]	h [kJ/kg]	s [kJ/(kg K)]
375	0.01056	2567.5	2752.3	5.494	0.00768	2449.1	2602.6	5.228
400	0.01246	2684.3	2902.4	5.721	0.00995	2617.9	2816.9	5.553
450	0.01520	2845.4	3111.4	6.021	0.01272	2807.2	3061.7	5.904
500	0.01739	2972.4	3276.7	6.242	0.01479	2945.3	3241.2	6.145
600	0.02107	3192.5	3561.3	6.589	0.01819	3175.3	3539.0	6.508
700	0.02434	3397.5	3823.5	6.873	0.02113	3385.1	3807.8	6.799
800	0.02741	3599.7	4079.3	7.124	0.02387	3590.1	4067.5	7.053

	$p = 250.0$ bar $= 25.0$ MPa				$p = 300.0$ bar $= 30.0$ MPa			
t [°C]	v [m^3/kg]	u [kJ/kg]	h [kJ/kg]	s [kJ/(kg K)]	v [m^3/kg]	u [kJ/kg]	h [kJ/kg]	s [kJ/(kg K)]
375	0.00196	1799.9	1849.4	4.034	0.00179	1738.1	1791.8	3.931
400	0.00601	2428.5	2578.6	5.140	0.00280	2068.9	2152.8	4.476
450	0.00918	2721.2	2950.6	5.676	0.00674	2618.9	2821.0	5.442
500	0.01114	2887.3	3165.9	5.964	0.00869	2824.0	3084.7	5.796
600	0.01414	3140.0	3493.5	6.364	0.01145	3103.4	3446.7	6.237
700	0.01664	3359.9	3776.0	6.670	0.01365	3334.3	3743.9	6.560
800	0.01892	3570.7	4043.8	6.932	0.01563	3551.2	4020.0	6.830

(*Continued*)

Table A.4. (*Continued*)

	$p = 400.0$ bar $= 40.0$ MPa			
t [°C]	v [m^3/kg]	u [kJ/kg]	h [kJ/kg]	s [kJ/(kg K)]
375	0.00164	1677	1742.6	3.829
400	0.00191	1854.9	1931.4	4.115
450	0.00369	2364.2	2511.8	4.945
500	0.00562	2681.6	2906.5	5.474
600	0.00809	3026.8	3350.4	6.017
700	0.00993	3282.0	3679.1	6.374
800	0.01152	3511.8	3972.6	6.661

Source: Adapted from "NISTIR 5078 — Thermodynamic Properties of Water: Tabulation from the IAPWS Formulation 1995 for the Thermodynamic Properties of Ordinary Water Substance for General and Scientific Use" available at https://nvlpubs.nist.gov/nistpubs/Legacy/IR/nistir5078.pdf (Accessed 31 July 2018). Internal energy was calculated as $u = h - pv$. See also the note for Table A.2.

Table A.5. Saturated refrigerant R 134a.

t [°C]	p [kPa]	Specific volume [m^3/kg] v_f	v_g	Specific enthalpy [kJ/kg] h_f	h_g	Specific entropy [kJ/(kg K)] s_f	s_g
−60	15.94	0.00067981	1.076	123.96	361.51	0.6871	1.8016
−55	21.86	0.00068629	0.8013	130.07	364.67	0.7154	1.7908
−50	29.48	0.00069295	0.6057	136.21	367.83	0.7432	1.7812
−45	39.114	0.00069979	0.4645	142.37	371.00	0.7705	1.7726
−40	51.22	0.00070681	0.3610	148.57	374.16	0.79731	1.7649
−35	66.14	0.00071408	0.2840	154.81	377.31	0.8238	1.7580
−30	84.36	0.00072155	0.2259	161.10	380.45	0.8498	1.7519
−25	106.37	0.00072929	0.1816	167.43	383.57	0.8755	1.7465
−20	132.68	0.00073735	0.1474	173.82	386.66	0.9009	1.7417
−15	163.87	0.00074571	0.1207	180.27	389.72	0.9261	1.7375
−10	200.52	0.00075438	0.09963	186.78	392.75	0.9509	1.7337
−5	243.24	0.00076348	0.08284	193.35	395.74	0.9756	1.7303
0	292.69	0.00077298	0.06935	200.00	398.68	1.0000	1.7274
5	349.54	0.00078302	0.05841	206.72	401.57	1.0242	1.7247
10	414.49	0.00079352	0.049476	213.53	404.40	1.0483	1.7224
15	488.25	0.00080463	0.042123	220.42	407.16	1.0722	1.7203
20	571.59	0.00081639	0.036028	227.40	409.84	1.0960	1.7183
25	665.26	0.00082898	0.030942	234.47	412.44	1.1197	1.7166
30	770.08	0.00084232	0.026671	241.65	414.94	1.1432	1.7149
35	886.85	0.00085675	0.023062	248.94	417.32	1.1668	1.7132
40	1016.5	0.00087222	0.019994	256.35	419.58	1.1903	1.7115
45	1159.8	0.00088885	0.017370	263.90	421.69	1.2138	1.7097
50	1317.7	0.00090744	0.015114	271.59	423.463	1.2373	1.7078
55	1491.3	0.00092764	0.013162	279.44	425.36	1.2609	1.7056

Source: Adapted from M.L. Huber and M.O. McLinden, "Thermodynamic Properties of R134a (1,1,1,2-tetrafluoroethane)" (1992). International Refrigeration and Air Conditioning Conference. Paper 184. See http://docs.lib.purdue.edu/iracc/184 (Accessed 31 July 2018).

Note: Reliable thermodynamic data for R134a can also be found at https://www.chemours.com/Refrigerants/en_US/assets/downloads/freon-134a-properties-uses-storage-handling.pdf (Accessed 31 July 2018).

Table A.6. Thermal conductivity of some materials at about 20 °C.

Material/ Substance	Thermal conductivity k [W/(m K)]	Material/ Substance	Thermal conductivity k [W/(m K)]
Air, atmosphere (gas)	0.0261	Limestone, coarse grain, dry	0.95
Aluminum (0.3% impurities)	228	Magnesium (Mg)	157
Aluminum (1% impurities)	199	Mercury, liquid (Hg)	8.3
Asphalt	0.692	Nylon 6*	0.21
Brass	103	Platinum (Pt)	70
Brick (masonry)	0.66–1.0	Polycarbonate*	0.19
Bronze (75 % Cu, 25 % Sn)	26	Polyester**	0.15–0.4
Concrete, dry	0.78	Polyethylene low density, PEL*	0.33
Concrete, reinforced	1.3	Polyethylene high density, PEH*	0.46 - 0.52
Copper (Cu)	388	Polypropylene, PP*	0.14
Carbon Steel	45	Polystyrene**	0.1–0.13
Cotton, fabric	0.08	Polyvinylchloride, PVC*	0.13–0.17
Fiberglass	0.04	Rubber, natural	0.15
Glass, window	0.78	Sand, dry	0.35
Glass, wool (fine)	0.052	Sand, moist (10 % H_2O)	1.0
Glass, wool (packed)	0.038	Silver (Ag)	419
Ground or soil, (30 % moisture by volume	2.42	Steel, Carbon 1%	45
Ground or soil, dry area	0.52	Stainless steel	14
Gypsum	0.43	Tin (Sn)	62
Hair felt	0.047	Vacuum	0
Ice (0 °C, 32 °F)	2.2	Wood across the grain (oak)	0.21
Iron (Fe)	62	Wood along the grain (oak)	0.35
Lead (Pb)	36	Wood across the grain (pine)	0.10
Limestone, fine grain, dry	0.69	Wood along the grain (pine)	0.24
Limestone, fine grain, 10% moisture	0.95	Zinc (Zn)	114

Source: [59] except for items marked (*) that are from https://www.m-ep.co.jp/en/pdf/product/iupi_nova/physicality_04.pdf, and marked (**) that are from https://www.professionalplastics.com/professionalplastics/ThermalPropertiesofPlasticMaterials.pdf (Accessed 1 August 2018).

Table A.7. Properties of dry air at 1 atm.

Temperature [°C]	Temperature [K]	Density ρ [kg/m^3]	Enthalpy h [kJ/kg]	Specific heat c_p [J/(kg K)]	Dynamic viscosity μ [10^{-6} Pa s]	Thermal conductivity k [W/(m K)]	Pr
−30	243.15	1.453	−30.163	1005.4	15.706	0.0217	0.728
−25	248.15	1.424	−25.135	1005.4	15.967	0.0221	0.726
−20	253.15	1.395	−20.108	1005.4	16.225	0.0225	0.725
−15	258.15	1.368	−15.081	1005.4	16.480	0.0229	0.723
−10	263.15	1.342	−10.055	1005.5	16.733	0.0233	0.722
−5	268.15	1.317	−5.028	1005.5	16.984	0.0237	0.721
0	273.15	1.293	0.000	1005.6	17.233	0.0241	0.720
5	278.15	1.270	5.028	1005.7	17.480	0.0245	0.719
10	283.15	1.247	10.058	1005.8	17.725	0.0248	0.718
15	288.15	1.225	15.089	1005.9	17.967	0.0252	0.717
20	293.15	1.204	20.122	1006.1	18.208	0.0256	0.716
25	298.15	1.184	25.157	1006.3	18.446	0.0260	0.715
30	303.15	1.164	30.194	1006.5	18.683	0.0263	0.714
35	308.15	1.146	35.234	1006.7	18.917	0.0267	0.713
40	313.15	1.127	40.277	1006.9	19.150	0.0271	0.712
45	318.15	1.109	45.323	1007.2	19.381	0.0274	0.711
50	323.15	1.092	50.372	1007.4	19.610	0.0278	0.711
55	328.15	1.076	55.426	1007.7	19.838	0.0282	0.710
60	333.15	1.059	60.483	1008.1	20.064	0.0285	0.709
65	338.15	1.044	65.545	1008.4	20.288	0.0289	0.709
70	343.15	1.028	70.612	1008.7	20.510	0.0292	0.708
75	348.15	1.014	75.684	1009.1	20.731	0.0296	0.707
80	353.15	0.999	80.761	1009.5	20.950	0.0299	0.707
85	358.15	0.985	85.844	1009.9	21.167	0.0303	0.706
90	363.15	0.972	90.933	1010.4	21.383	0.0306	0.706
95	368.15	0.958	96.027	1010.8	21.598	0.0310	0.705
100	373.15	0.945	101.129	1011.3	21.811	0.0313	0.705
105	378.15	0.933	106.237	1011.8	22.022	0.0316	0.705
110	383.15	0.921	111.353	1012.3	22.232	0.0320	0.704
115	388.15	0.909	116.476	1012.8	22.441	0.0323	0.704
120	393.15	0.897	121.606	1013.4	22.648	0.0326	0.703
125	398.15	0.886	126.745	1014.0	22.854	0.0330	0.703
130	403.15	0.875	131.892	1014.6	23.059	0.0333	0.703
135	408.15	0.864	137.047	1015.2	23.262	0.0336	0.703
140	413.15	0.854	142.211	1015.8	23.464	0.0339	0.702
145	418.15	0.843	147.385	1016.4	23.664	0.0343	0.702

Source: The values of ρ, μ, k and c_p were calculated with the formulas of approximation from F. J. McQuillan, J. R. Culham and M. M. Yovanovich, *Properties of Dry Air at One Atmosphere*, UW/MHTL 8406, G-01, Microelectronics Heat Transfer Lab, University of Waterloo, Waterloo, Ontario, June 1984, the values of h and Pr were calculated using their equations of definition.

Table A.8. Properties of water at 1 atm.

Temp. t [°C]	Density ρ [kg/m^3]	Enthalpy h [kJ/kg]	Specific heat c_p [kJ/(kg K)]	Dynamic viscosity μ [mPa s]	Thermal conductivity k [W/(m K)]	Pr
0.01	999.80	0.01367	4.2174	1.792	0.565	13.37
5	999.91	21.007	4.2012	1.519	0.575	11.11
10	999.64	42.002	4.1921	1.307	0.584	9.38
15	999.04	62.963	4.1863	1.138	0.594	8.03
20	998.15	83.898	4.1826	1.002	0.602	6.96
25	997.00	104.814	4.1803	0.890	0.610	6.10
30	995.61	125.717	4.1790	0.797	0.618	5.39
35	994.00	146.613	4.1786	0.719	0.626	4.81
40	992.18	167.505	4.1788	0.653	0.633	4.32
45	990.18	188.398	4.1796	0.596	0.639	3.91
50	988.00	209.296	4.1809	0.547	0.646	3.56
55	985.66	230.201	4.1825	0.504	0.652	3.25
60	983.16	251.116	4.1846	0.466	0.657	2.99
65	980.51	272.044	4.1871	0.433	0.663	2.76
70	977.73	292.987	4.1900	0.404	0.668	2.56
75	974.80	313.948	4.1932	0.378	0.672	2.39
80	971.75	334.927	4.1969	0.354	0.677	2.23
85	968.58	355.927	4.2010	0.333	0.681	2.09
90	965.28	376.950	4.2056	0.315	0.685	1.97
95	961.87	397.998	4.2107	0.297	0.689	1.86

Source: Properties were calculated with the formulae of approximation from F. P. Incropera and D. P. DeWitt *Introduction to Heat Transfer*, 3rd edition, John Wiley & Sons, Inc., New York, USA, 1996.

Table A.9. Emissivity factors for some materials.

Surface material	Temperature [K]	Emissivity ε	Surface material	Temperature [K]	Emissivity ε
Aluminum, polished	300	0.04	Nickel, polished*	373	0.072
Aluminum, oxidized	300	0.09	Nickel oxide*	922	0.59
Aluminum oxide*	550	0.63	Oak, planed*	294	0.90
Brass, highly polished*	520	0.028	Paint, aluminum*	373	0.52
	630	0.031	Paint, oil (all colors)*	373	0.92–0.96
Building brick	300	0.93	Paper*	292	0.924
Chromium, polished	300	0.08	Plaster	300	0.92
Copper, polished	300	0.02	Quartz glass	300	0.93
Copper, oxidized	300	0.56	Roofing paper*	294	0.91
Glass, smooth*	295	0.94	Rubber (hard, glossy)*	296	0.94
Iron, oxidized*	373	0.74	Silver, polished	300	0.02
Iron, tin-plated*	373	0.07	Steel, polished	300	0.07
Iron oxide*	772	0.85	Steel, oxidized	300	0.79
Lampblack paint	300	0.96	Tungsten, polished	300	0.04
Lead, unoxidized*	400	0.057	Tungsten, filament	300	0.32
Mercury, liquid	300	0.1	Water (and ice)	273 (373)	0.96

Source: [1] except for items marked (*) that are from [59].

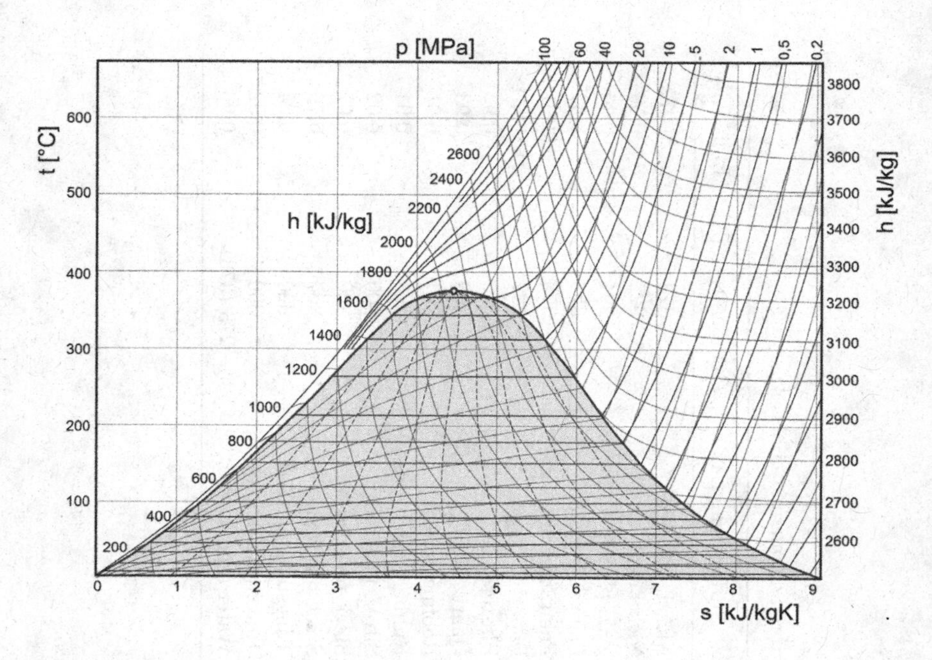

Fig. A.1. $T - s$ diagram for water.

Source: *Water/Steam $T - s$ (temperature–entropy) diagram* by Kaboldy [CC BY-SA 3.0 (https://creativecommons.org/licenses/by-sa/3.0)], from Wikimedia Commons. https://commons.wikimedia.org/wiki/File:T-s_diagram.svg (Accessed 1 August 2018).

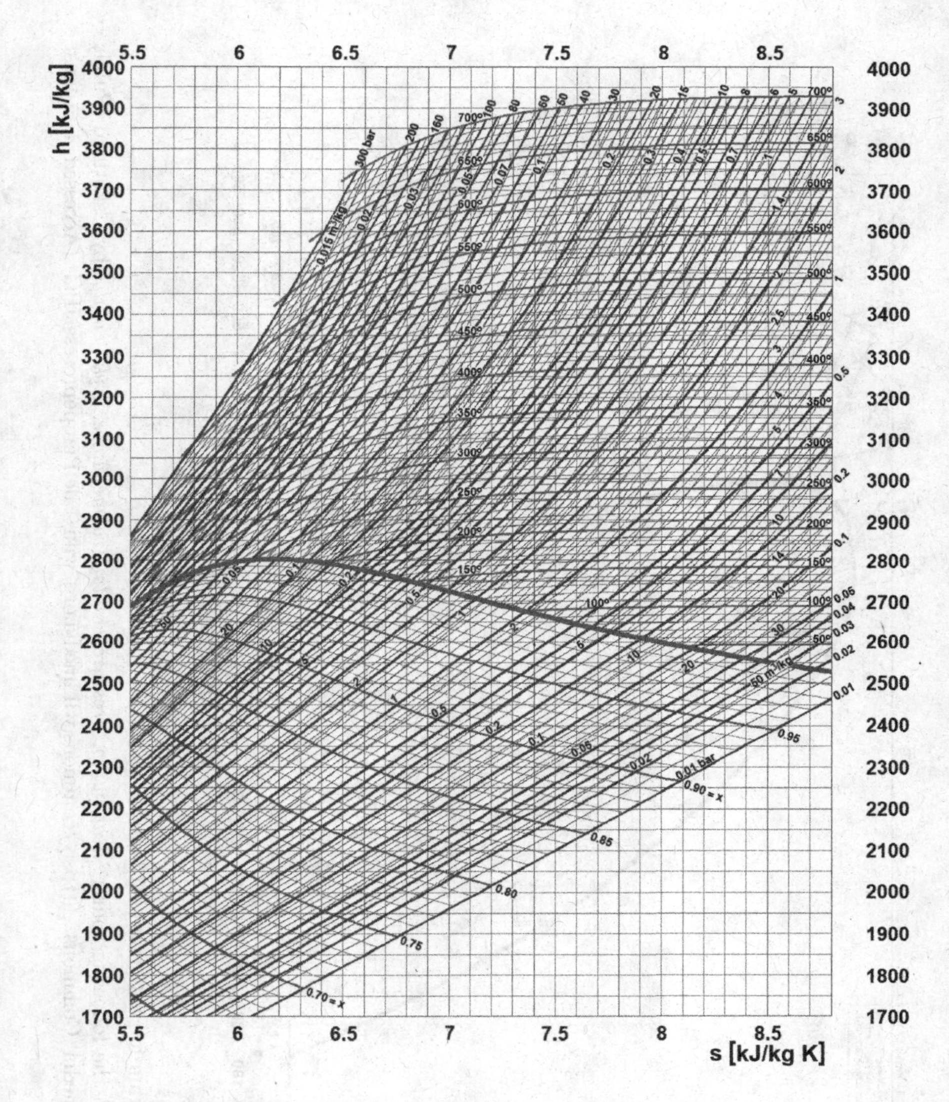

Fig. A.2. Mollier ($h - s$) diagram for water.

Source: Adapted from the following source: *H*, *s-diagramm* by User : Pretenderrs [GFDL (http://www.gnu.org/copyleft/fdl.html) or CC BY-SA 3.0 (https://creativecommons.org/licenses/by-sa/3.0)], from Wikimedia Commons. https://commons.wikimedia.org/wiki/File:H,_s-diagramm.PNG (Accessed 1 August 2018).

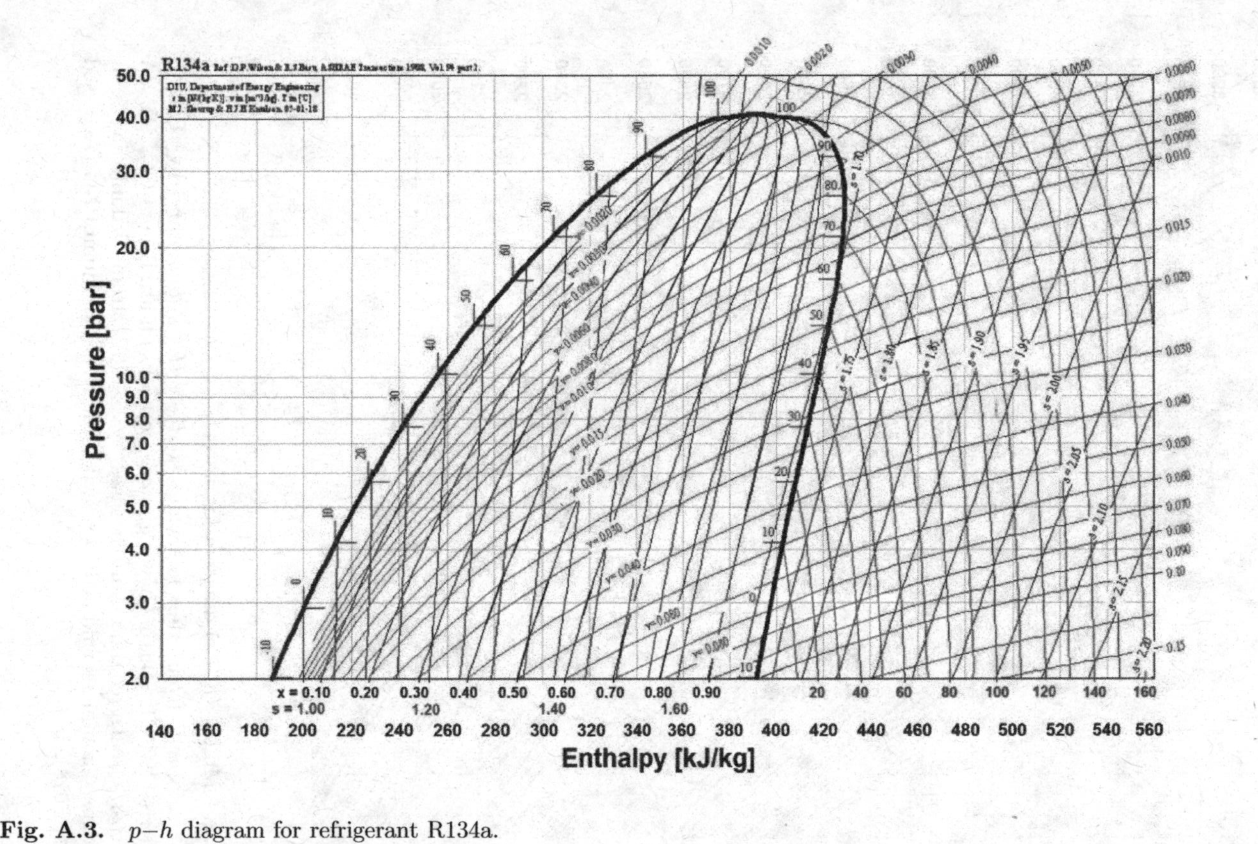

Fig. A.3. p–h diagram for refrigerant R134a.

Source: Adapted from the following source: *heat pump*, *pressure entalpy*, *varmepumpe*, *trykk entalpi* by Cskotland [Public domain], from Wikimedia Commons. https://commons.wikimedia.org/wiki/File:Pumpeprosess.JPG (Accessed 1 August 2018).

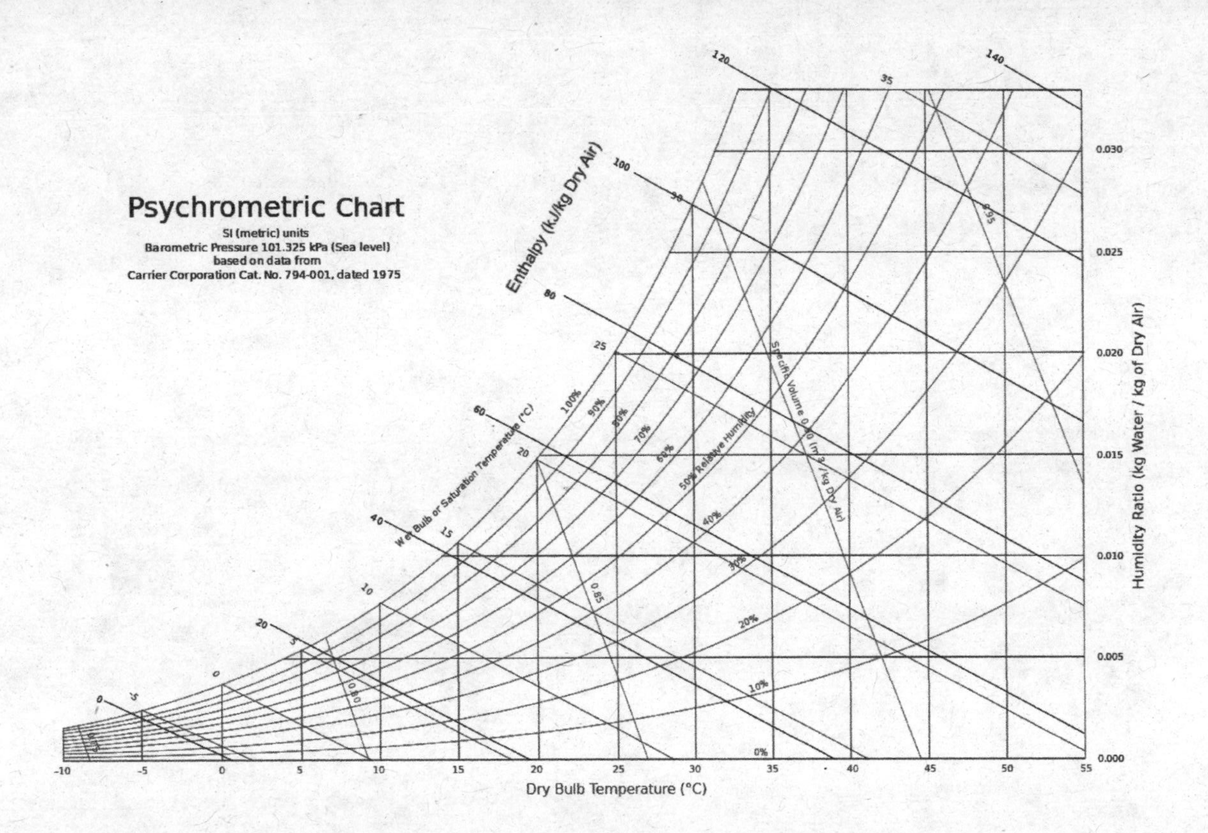

Fig. A.4. Psychrometric chart.

Source: Adapted from the following source: *Psychrometric Chart for Sea-level Pressure in SI units* by Arthur Ogawa [GFDL (http://www.gnu.org/copyleft/fdl.html) or CC BY-SA 3.0 (https://creativecommons.org/licenses/by-sa/3.0)], from Wikimedia Commons. https://commons.wikimedia.org/wiki/File:Psychrometric.SeaLevel.SI.svg (Accessed 1 August 2018).

Solutions to Problems

Note: The exact conversion of temperature between degrees Celsius and kelvin is as follows:

$$T(\mathrm{K}) = t(^\circ\mathrm{C}) + 273.15$$

Often, the rounded value of 273 is used instead of 273.15. All examples to follow will use 273.

1.1. The gauge pressure is the pressure indicated by the manometer (above the atmospheric level):

$$p_g = 2.8 \text{ bar} = 2.8 \times 10^5 \text{ Pa}$$

Considering the ambient pressure $p_{\mathrm{atm}} = 10^5$ Pa, the absolute pressure is

$$p = p_g + p_{\mathrm{atm}} = 2.8 \times 10^5 + 10^5 = 3.8 \times 10^5 \text{ Pa}$$

1.2. The atmospheric pressure is

$$750 \text{ torr} = 750 \times 133.3 = 99\,975 \text{ Pa} = 0.99975 \text{ bar} \cong 1 \text{ bar}$$

The gauge reading will be $p_g = p - p_{\mathrm{atm}} = 16 - 1 = 15$ bar

1.3. (c)

1.4. (b)

1.5. (c)

1.6. (a)

2.1. If pressure varies, the saturation temperature of the liquid will change also, as represented qualitatively in Fig. 2.3 (the vapor pressure curve). Therefore, if pressure increases during the process, the saturation (boiling) temperature will increase also, and vice versa.

2.2. According to the definition of steam quality,

$$x = \frac{m_{\text{sat.vapor}}}{m_{\text{sat.liquid}} + m_{\text{sat.vapor}}} = \frac{m_g}{m_f + m_g} = \frac{m_g}{m}$$

$$m = m_{\text{sat.liquid}} + m_{\text{sat.vapor}} = m_f + m_g = 0.15\,\text{kg} + 1\,\text{kg} = 1.15\,\text{kg}$$

Therefore

$$x = \frac{m_g}{m} = \frac{1}{1.15} = 0.8696 \cong 0.87 = 87\ \%$$

2.3. The air inside the container is considered an ideal gas. Therefore the ideal gas equation of state can be applied.

(a) The air is trapped inside a container with rigid walls. It constitutes a closed system and a rigid one, because it does not allow any work exchange with the surroundings. Heat can be transferred, however, because any metallic material is conductive.

(b) The specific volume v is

$$v = \frac{V}{m} = \frac{10 \times 10^{-3}}{2} = 5 \times 10^{-3}\text{m}^3/\text{kg} = 0.005\,\text{m}^3/\text{kg}$$

(c) The absolute pressure can be obtained from the equation of state:

$$p = \frac{RT}{v} = \frac{287 \times (273 + 24)}{0.005} = 170.5 \times 10^5\,\text{Pa} = 170.5\,\text{bar}$$

2.4. *Assumption*: The gas is considered an ideal gas. The ideal gas equation of state can be applied.

Converting all quantities to SI units, one obtains

$$p = 4\,\text{kgf/cm}^2 = 4 \times \frac{9.81}{10^{-4}}\frac{\text{N}}{\text{m}^2} = 392\,400\,\text{Pa}$$

$$V = 9\,\text{L} = 9 \times 10^{-3}\,\text{m}^3$$

$$\dot{m} = 0.3\,\text{kg/min} = \frac{0.3}{60}\,\text{kg/s} = 5 \times 10^{-3}\text{kg/s}$$

The mass of gas inside the container at the valve closure is

$$m = \dot{m} \times \tau = 5 \times 10^{-3} \times 6 = 30 \times 10^{-3}\,\text{kg}$$

The equilibrium temperature is

$$T = \frac{pV}{mR} = \frac{392.4 \times 10^3 \times 9 \times 10^{-3}}{30 \times 10^{-3} \times 297} = 396.4\,\text{K}$$

2.5. *Assumption*: Helium is considered an ideal gas. The ideal gas equation of state can be applied.

(a) A closed system; no mass can flow into or out of the system.
(b)

$$v = \frac{RT}{p} = \frac{2077 \times (273 + 18)}{2.4 \times 10^5} = 2.518\,\text{m}^3/\text{kg}$$

3.1. *Assumption*: The air is considered an ideal gas. The non-flow energy equation can be used.

For the first expansion:

$$Q - W = \Delta U; \quad \Delta U = 18 - 25 = -7\,\text{kJ}$$

The second expansion occurs between the same states, therefore the change in internal energy (a state function) is the same as in the first case. This time, $W = Q - \Delta U = 10 - (-7) = 17\,\text{kJ}$.

3.2. *Assumption*: The air is considered an ideal gas. The ideal gas equation of state can be applied.

$$h = u + pv = u + RT$$

$$h = 323\,000 + 287 \times (273 + 177) = 452\,150\,\text{J/kg} = 452.2\,\text{kJ/kg}$$

3.3. *Assumption*: The system considered is a closed one.

According to the non-flow energy equation, $Q - W = \Delta U$. If $Q < W$, the work produced by the system during the process will utilize the entire heat input and the balance will be covered by the reduction of internal energy. Therefore $\Delta U = Q - W = 630 - 915 = -285\,\text{J}$. If work is done by the surroundings on the system during the process, $\Delta U = Q - W = 630 - (-915) = 1545\,\text{J}$; the internal energy of the system increases.

3.4. (a)

3.5. (b)

4.1. *Assumptions*: The gas is considered an ideal gas. The ideal gas equation of state can be applied. The atmospheric pressure is $p_{\text{atm}} = 101325$ Pa $= 1.01$ bar.

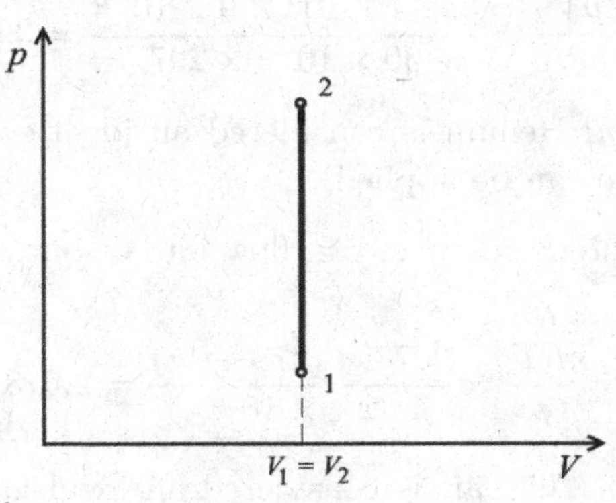

The air inside the tires (closed system) undergoes an isochoric process (V = cst). State 1 — start of the trip; state 2 — end of the trip. Therefore

$$\frac{p_1}{T_1} = \frac{p_2}{T_2}$$

$$T_2 = T_1 \frac{p_{g2} + p_{\text{atm}}}{p_{g1} + p_{\text{atm}}} = 284 \frac{3.01 + 1.01}{2.81 + 1.01} = 298.9\,\text{K}$$

4.2. *Assumption*: The system (ideal gas) is a closed one.

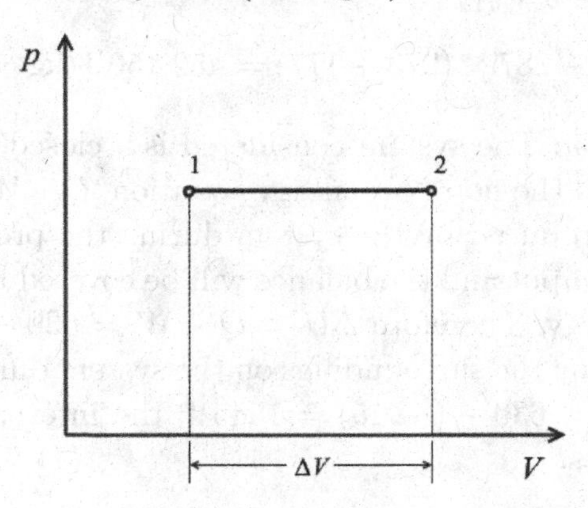

The process is isobaric and according to the non-flow energy equation

$$Q - W = \Delta U \Rightarrow W = Q - \Delta U = p\Delta V$$

$$\Delta V = \frac{Q - \Delta U}{p} = \frac{1350 - 1150}{1.01 \times 10^5} = 0.00198\,\text{m}^3$$

The volume increases by 1.98×10^{-3} m^3 (positive work).

4.3. *Assumption*: The system (ideal gas) is a closed one.
During an isothermal process $Q = W$. Therefore $W = 5200$ J.

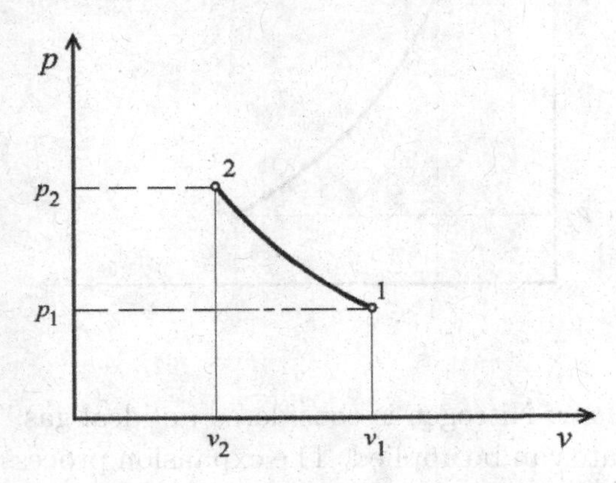

4.4. *Assumptions*: The air is considered an ideal gas. The ideal gas equation of state can be applied. $R = 287$ J/(kg K). The process is slow enough for the system to be always in thermal equilibrium with the surrounding (isothermal process):

(a) $p_1 v_1 = p_2 v_2 = RT$

$$T = \frac{p_1\, v_1}{R} = \frac{1.2 \times 10^5 \times 1.25}{287} = 522.6\,\text{K}$$

(b) $w_{12} = p_1\, v_1 \ln \dfrac{p_1}{p_2} = 1.2 \times 10^5 \times 1.25 \times \ln \dfrac{1.2}{5} = -214\,067\,\text{J/kg}$

The work is negative because it is done by the surroundings on the system (compression).

(c) For an isothermal process $q_{12} = w_{12} = -214\ 067$ J/kg. Heat is evacuated from the system (negative).

(d) The change in enthalpy, as the change in internal energy, is zero:

$$\Delta h = h_2 - h_1 = (u_2 + p_2\, v_2) - (u_1 + p_1\, v_1)$$

$$u_2 = u_1; \quad p_1 v_1 = p_2 v_2 \Rightarrow \Delta h = 0$$

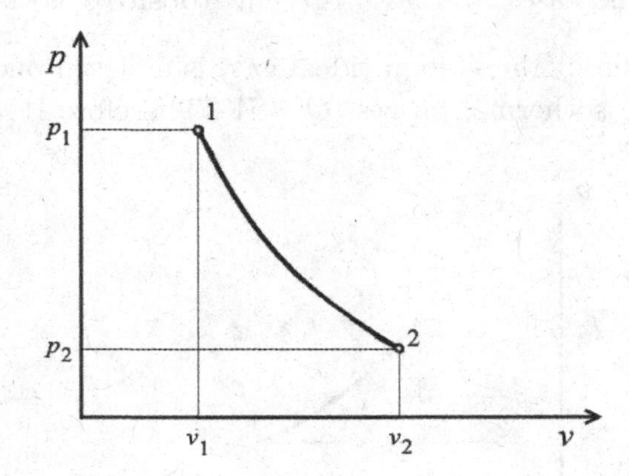

4.5. *Assumptions*: Nitrogen is considered an ideal gas. The ideal gas equation of state can be applied. The expansion process is polytropic ($n = 1.5$).

$$p\, v^n = \text{cst}; \quad \Rightarrow \; p_1 v_1^n = p_2 v_2^n$$

The work associated to the process is

$$w = \frac{p_2 v_2 - p_1 v_1}{1 - n} = \frac{p_1 v_1}{n - 1}\left[1 - \left(\frac{v_1}{v_2}\right)^{n-1}\right]$$

$$= \frac{6 \times 10^5 \times 0.625}{1.5 - 1}\left[1 - \left(\frac{1}{2.5}\right)^{1.5-1}\right] = 110\ 263 \text{ J/kg}$$

5.1. *Assumptions*: The air is considered an ideal gas ($k = 1.4$). The ideal gas equation of state can be applied. $R = 287$ J/(kg K). The compression process is adiabatic (ideal compressor, open system).

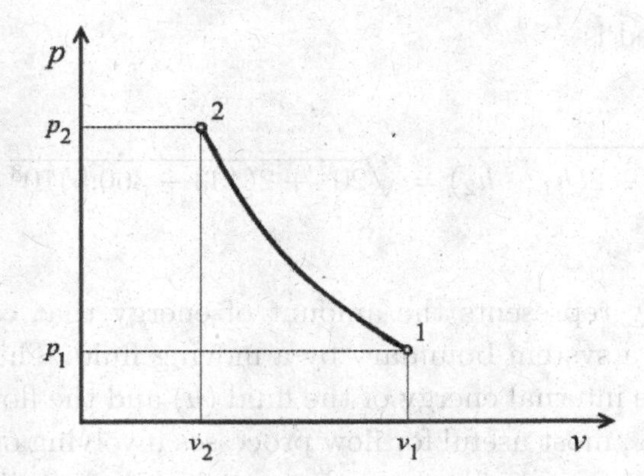

The atmospheric pressure is considered $p_1 = 101\ 325$ Pa:

$$p_1 \dot{V}_1 = \dot{m} R T_1$$

$$\dot{m} = \frac{p_1 \dot{V}_1}{RT_1} = \frac{101\ 325 \times 50}{287 \times (273 + 15)} = 61.29 \text{ kg/min} = 1.02 \text{ kg/s}$$

$$w = \Delta h = \frac{kRT_1}{k-1}\left[1 - \left(\frac{p_2}{p_1}\right)^{\frac{k-1}{k}}\right]$$

$$= \frac{1.4 \times 287 \times (273 + 15)}{1.4 - 1}\left[1 - \left(\frac{220}{1.01}\right)^{0.4/1.4}\right] = -1058 \text{ kJ/kg}$$

The work is negative (compression). The power consumed by the ideal compressor is:

$$P = w\,\dot{m} = -1058 \times 1.02 = -1079 \text{ kW}$$

5.2. *Assumption*: The combustion products are considered an ideal gas ($k = 1.38$). The ideal gas equation of state can be applied. The expansion process $1 - 2$ is adiabatic reversible (quasistatic):

$$\frac{T_2}{T_1} = \left(\frac{p_2}{p_1}\right)^{\frac{k-1}{k}}$$

The exit temperature T_2 is

$$T_2 = T_1 \left(\frac{p_2}{p_1}\right)^{\frac{k-1}{k}} = 433 \left(\frac{100}{220}\right)^{\frac{1.38-1}{1.38}} = 348.5 \text{ K}$$

The exit speed is

$$V_2 = \sqrt{V_1^2 + 2(h_1 - h_2)} = \sqrt{20^2 + 2(433 - 360.5)10^3} = 381 \text{ m/s}$$

5.3. Enthalpy represents the amount of energy that can be transferred across a system boundary by a moving fluid. This energy has two parts: the internal energy of the fluid (u) and the flow work (pv). It is, therefore, most useful for flow processes involving open systems. However, being a state function, the enthalpy characterizes the state of any system, open or not.

6.1. *Assumption*: The air is considered an ideal gas. The ideal gas equation of state can be applied.

Initial data: $p_1 = 1$ bar $= 10^5$ Pa; $V_1 = 1.5$ m^3; $T_1 = 273 + 55 = 328$ K; $p_2 = 5$ bar; $t_3 = 1750$ °C; $V_4 = 2.5$ m^3; $c_p = 1000$ J/(kg K); $R = 287$ J/(kg K).

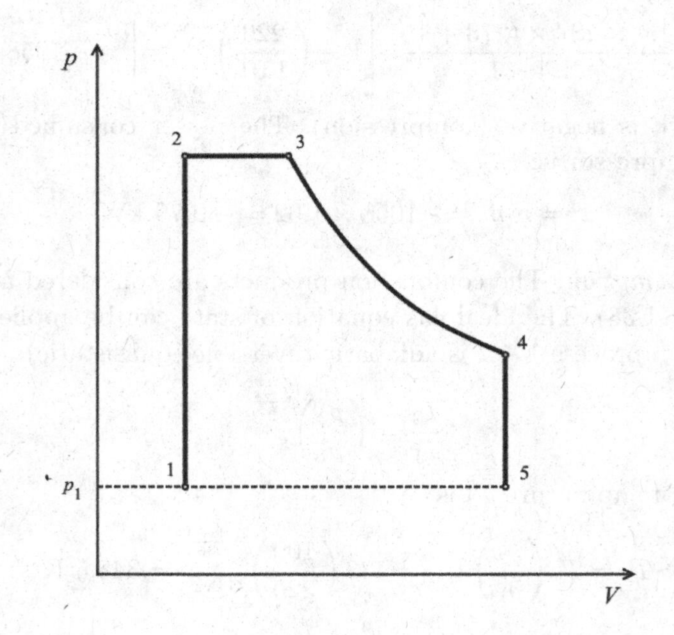

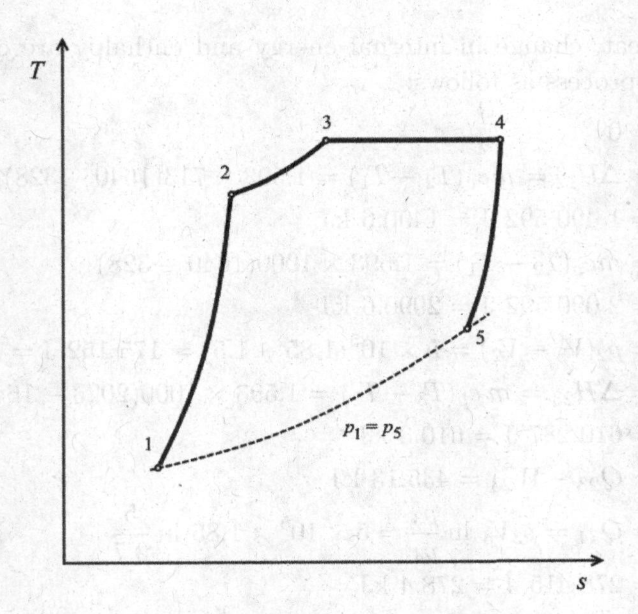

(a) For the process 1–2: $V_2 = V_1 = 1.5\ \text{m}^3$;

$$T_2 = T_1 \frac{p_2}{p_1} = 328 \frac{5 \times 10^5}{1 \times 10^5} = 1640\ \text{K}$$

For the process 2–3: $p_3 = p_2 = 5$ bar;

$$V_3 = V_2 \frac{T_3}{T_2} = 1.5 \frac{273 + 1750}{1640} = 1.85\ \text{m}^3$$

For the process 3–4: $T_4 = T_3 = 2023$ K

$$p_4 = p_3 \frac{V_3}{V_4} = 5 \frac{1.85}{2.5} = 3.7\ \text{bar}$$

For the process 4–5: $p_5 = p_1 = 1$ bar; $V_5 = V_4 = 2.5\ \text{m}^3$;

$$T_5 = T_4 \frac{p_5}{p_4} = 2023 \frac{1}{3.7} = 546.7\ \text{K}$$

(b) $c_v = c_p - R = 713\ \text{J/(kg K)}$;

$$m = \frac{p_1 V_1}{R T_1} = \frac{10^5 \times 1.5}{287 \times 328} = 1.593\ \text{kg}$$

Work, heat, change in internal energy and enthalpy are calculated for each process as follows:

$$
\begin{aligned}
W_{12} &= 0 \\
Q_{12} &= \Delta U_{12} = mc_v(T_2 - T_1) = 1.593 \times 713(1640 - 328) \\
&= 1\,490\,592 \text{ J} = 1490.6 \text{ kJ} \\
\Delta H_{12} &= mc_p(T_2 - T_1) = 1.593 \times 1000(1640 - 328) \\
&= 2\,090\,592 \text{ J} = 2090.6 \text{ kJ} \\
W_{23} &= p_2(V_3 - V_2) = 5 \times 10^5(1.85 - 1.5) = 175\,152 \text{ J} = 175.2 \text{ kJ} \\
Q_{23} &= \Delta H_{23} = m\,c_p(T_3 - T_2) = 1.593 \times 1000(2023 - 1640) \\
&= 610\,287 \text{ J} = 610.3 \text{ kJ} \\
\Delta U_{23} &= Q_{23} - W_{23} = 435.13 \text{ kJ} \\
W_{34} &= Q_{34} = p_3 V_3 \ln \frac{p_3}{p_4} = 5 \times 10^5 \times 1.85 \ln \frac{5}{3.7} \\
&= 278\,415 \text{ J} = 278.4 \text{ kJ} \\
\Delta U_{34} &= m\,c_v\,\Delta T = 0, \quad \Delta H_{34} = m\,c_p\,\Delta T = 0 \\
W_{45} &= 0 \\
Q_{45} &= \Delta U_{45} = m\,c_v(T_5 - T_4) = 1.593 \times 713(546.7 - 2023) \\
&= -1\,677\,295 \text{ J} = -1677.3 \text{ kJ} \\
\Delta H_{45} &= m\,c_p(T_5 - T_4) = 1.593 \times 1000(546.7 - 2023) \\
&= -2\,352\,448 \text{ J} = -2352.4 \text{ kJ}
\end{aligned}
$$

The total energy exchange with the surroundings is

$$W_{\text{tot}} = W_{12} + W_{23} + W_{34} + W_{45} = 453.6 \text{ kJ}$$

$$Q_{\text{tot}} = Q_{12} + Q_{23} + Q_{34} + Q_{45} = 702.0 \text{ kJ}$$

The total change in internal energy and enthalpy are

$$\Delta U_{\text{tot}} = \Delta U_{12} + \Delta U_{23} + \Delta U_{34} + \Delta U_{45} = 248.4 \text{ kJ}$$

$$\Delta H_{\text{tot}} = \Delta H_{12} + \Delta H_{23} + \Delta H_{34} + \Delta H_{45} = 348.4 \text{ kJ}$$

6.2. (a)

6.3. (b)

6.4. (a) and (c)

6.5. (c)

6.6. (b)

6.7. *Assumption*: CO_2 is considered a perfect gas (i.e., ideal gas with constant specific heats). The ideal gas equation of state can be applied.

$$\Delta S = m\, c_v \ln\left(\frac{T_2}{T_1}\right)$$

$$c_v = c_p - R = 825 - 189 = 636 \text{ J/(kg K)}$$

$$V = cst \Rightarrow \frac{T_2}{T_1} = \frac{p_2}{p_1} = 1.8$$

$$\Delta S = 1.5 \times 636 \times \ln 1.8 = 560.7 \text{ J/K}$$

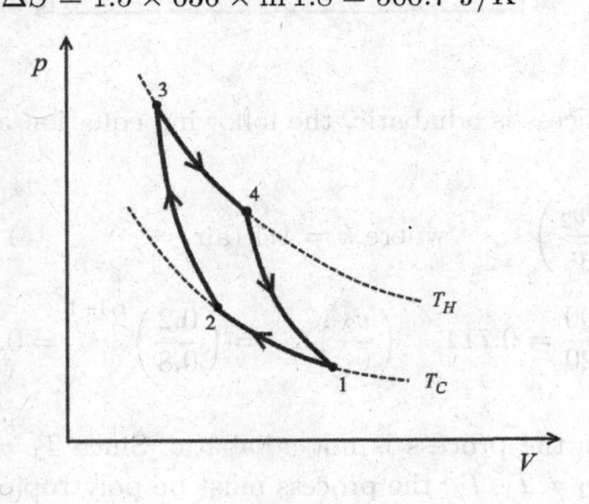

6.8. *Assumption*: The working fluid is an ideal gas. All processes are totally reversible:

$$\eta_{\text{th},C} = 1 - \frac{|Q_{\text{out}}|}{Q_{\text{in}}}$$

$$|Q_{\text{out}}| = |W_{\text{iso.in}}|, \quad Q_{\text{in}} = W_{\text{iso.out}}$$

$$\eta_{\text{th},C} = 1 - \frac{|W_{\text{iso.in}}|}{W_{\text{iso.out}}}$$

$$|W_{\text{iso.in}}| = (1 - \eta_{\text{th},C})\, W_{\text{iso.out}} = (1 - 0.2)\, 100 = 80 \text{ J}$$

Compression work is negative, therefore $W_{\text{iso.in}} = -80$ J

6.9. *Assumption*: The working fluid is an ideal gas. $R = 287$ J/(kg K); $c_p \approx 1000$ J/(kg K).

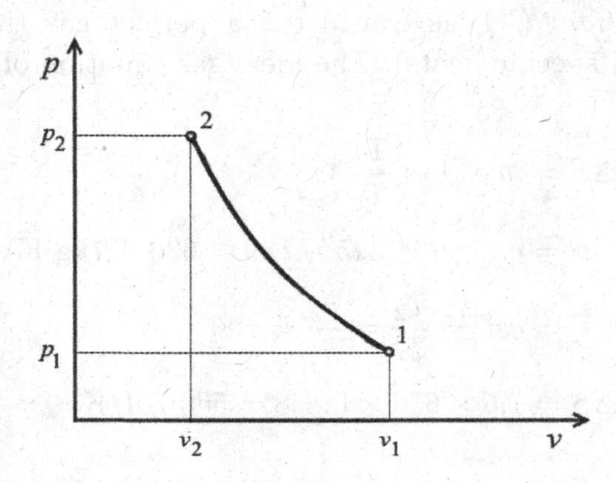

(a) If the process is adiabatic, the following equation applies:

$$\frac{T_1}{T_2} = \left(\frac{v_2}{v_1}\right)^{k-1} \qquad \text{where } k = 1.4 \text{ (air)}$$

$$\frac{T_1}{T_2} = \frac{300}{420} = 0.714, \quad \left(\frac{v_2}{v_1}\right)^{k-1} = \left(\frac{0.2}{0.8}\right)^{1.4-1} = 0.144 \neq 0.714$$

Therefore the process is not adiabatic. Since $T_1 \neq T_2, v_1 \neq v_2$ and $v_1/v_2 \neq T_1/T_2$, the process must be polytropic.

(b) So

$$\frac{T_1}{T_2} = \left(\frac{v_2}{v_1}\right)^{n-1} \Rightarrow \ln\left(\frac{T_1}{T_2}\right) = (n-1)\ln\left(\frac{v_2}{v_1}\right)$$

$$n = 1 + \ln\left(\frac{T_1}{T_2}\right) \Big/ \ln\left(\frac{v_2}{v_1}\right)$$

$$= 1 + \ln\left(\frac{360}{420}\right) \Big/ \ln\left(\frac{0.2}{0.8}\right) \cong 1.24$$

Notice that $1 < n < k$. In this case, heat flow and work flow are in opposite directions (e.g., if work is introduced, heat is evacuated from the system). If $n > k$, heat flow and work flow are in the same direction.

Polytropic work:

$$W_{12} = \frac{mRT_1}{n-1}\left[1-\left(\frac{v_1}{v_2}\right)^{n-1}\right]$$

$$= \frac{1 \times 287 \times 300}{1.24-1}\left[1-\left(\frac{0.8}{0.2}\right)^{1.24-1}\right] = -141\ 896 \text{ J}$$

Notice that the work is negative (compression):

$$c_v = c_p - R = 1000 - 287 = 713 \text{ J/(kg K)}$$

$$W_{12} - (W_{12}) = \Delta U$$

$$Q_{12} - (W_{12}) = mc_v(T_2 - T_1)$$

$$Q_{12} = mc_v(T_2 - T_1) + (W_{12})$$

$$= 1 \times 713(420 - 300) + (-141896) = -56\ 336 \text{ J}$$

Heat is negative, therefore evacuated from the system. This confirms the conclusion based on the value of the polytropic exponent n.

7.1. *Assumption*: The two components are considered ideal gases. The ideal gas equation of state applies.

The two components are 1 and 2. Properties of the mixture will take no indices.

$$p_1 = p_2 = p; \quad T_1 = T_2 = T; \quad m_1 = m_2 = m_c;$$

$$p\,V_1 = \frac{m_c}{M_1}\bar{R}T; \quad pV_2 = \frac{m_c}{M_1}\bar{R}T$$

$$p\,M_1 = \rho_1\,\bar{R}T; \quad p\,M_2 = \rho_2\,\bar{R}T$$

$$M_1 = \frac{\rho_1\,RT}{p}; \quad M_2 = \frac{\rho_2\,RT}{p}$$

For the mixture:

$$pV = \left(\frac{m_c}{M_1} + \frac{m_c}{M_2}\right)\bar{R}T \quad \text{or} \quad pV = m_c\bar{R}T\left(\frac{1}{M_1} + \frac{1}{M_2}\right)$$

$$m_c = \frac{pV}{\bar{R}T\left(\frac{1}{M_1} + \frac{1}{M_2}\right)} \Rightarrow \rho = \frac{2m_c}{V} = \frac{2pV}{\bar{R}T\left(\frac{1}{M_1} + \frac{1}{M_2}\right)V}$$

$$\rho = \frac{2p}{\bar{R}T\left(\frac{p}{\rho_1 \bar{R}T} + \frac{p}{\rho_2 \bar{R}T}\right)} = \frac{2}{\frac{1}{\rho_1} + \frac{1}{\rho_2}}$$

$$= \frac{2\,\rho_1\,\rho_2}{\rho_1 + \rho_2} = \frac{2 \times 1.1 \times 0.8}{1.1 + 0.8} = 0.926 \text{ kg/m}^3$$

7.2. *Assumption*: The two constituents are considered ideal gases. The ideal gas equation of state applies.

(a) The mass of nitrogen:

$$m_{N_2} = \frac{p_{N_2} V_{N_2}}{R_{N_2} T_{N_2}} = \frac{101\,325 \times 0.79}{297 \times 273} = 0.9872 \text{ kg}$$

The mass of the mixture is

$$m = m_{N_2} + m_{\text{vap}} = 0.9872 + 0.09 = 1.0772 \text{ kg}$$

The mass fraction of nitrogen is

$$\frac{m_{N_2}}{m} = \frac{0.9872}{1.0772} = 0.9164$$

The mass fraction of water vapor is

$$\frac{m_{\text{vap}}}{m} = \frac{0.09}{0.0772} = 0.0835$$

(b) The specific gas constant for the mixture is

$$R = \frac{m_{N_2}}{m} R_{N_2} + \frac{m_{\text{vap}}}{m} R_{\text{vap}}$$

$$R = 0.9164 \times 297 + 0.0835 \times \frac{8314}{18} = 310.8 \text{ J/(kg K)}$$

(c) Partial pressures are

$$p_{N_2} = \frac{m_{N_2} R_{N_2} T}{V} = \frac{0.9872 \times 297 \times (273 + 25)}{22 \times 10^{-3}}$$

$$= 3971684 \text{ Pa} = 39.72 \text{ bar}$$

$$p_{\text{vap}} = \frac{m_{\text{vap}} R_{\text{vap}} T}{V} = \frac{0.09 \times 461.9 \times 298}{22 \times 10^{-3}}$$

$$= 563\,084.5 \text{ Pa} = 5.63 \text{ bar}$$

7.3. (b)

7.4. (b) and (c)

7.5. (b)

7.6. *Assumption*: Humid air is considered an ideal gas.

According to Dalton's law,

$$p = p_a + p_v$$

The partial pressure of dry air is

$$p_a = p - p_v = 101 \times 10^3 - 51 \times 10^2 = 95900 \text{ Pa} = 0.959 \text{ bar}$$

The absolute humidity is

$$\omega = 0.622 \frac{p_v}{p - p_v} = 0.622 \frac{5100}{101\,000 - 5100} = 0.033 \text{ kg}_v/\text{kg}_a$$

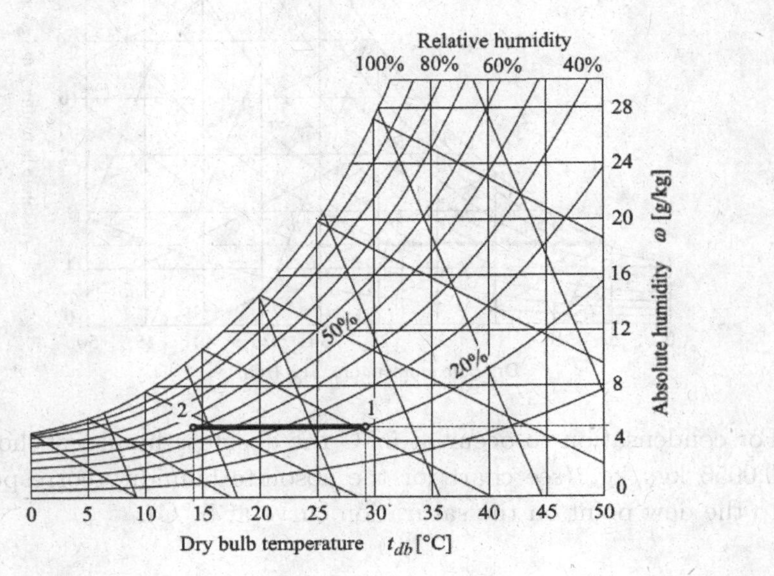

7.7. *Assumption*: Humid air is considered an ideal gas. The process is isobaric, occurring at atmospheric pressure. The psychrometric chart can be used.

The absolute humidity is constant during cooling. Its value can be read from the psychrometric chart:

$$t_1 = t_{db1} = 29\ ^\circ\text{C}, \quad \phi_1 = 20\ \% \Rightarrow \omega_1 = \omega_2 = 0.005\ \text{kg}_\text{v}/\text{kg}_\text{a}$$

By cooling the air, its relative humidity increases. From the psychrometric chart:

$$\omega_2 = 0.005\ \text{kg}_\text{v}/\text{kg}_\text{a}; \quad t_2 = t_{db2} = 14\ ^\circ\text{C} \Rightarrow \phi_2 = 50\ \%$$

7.8. *Assumption*: Humid air is considered an ideal gas. All processes occur at constant atmospheric pressure. This pressure is close enough to the normal atmospheric pressure (760 mmHg), so the standard psychrometric chart can be used.

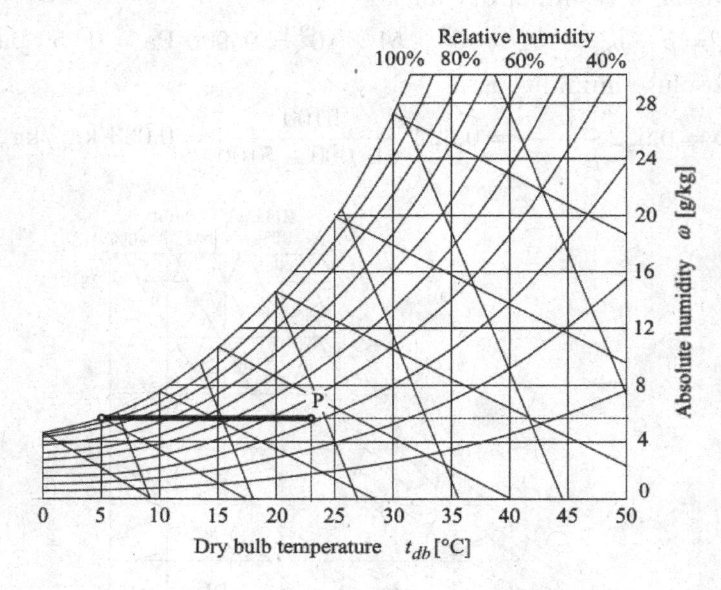

For condensation to occur at 5 °C the absolute humidity should be 0.0055 $\text{kg}_\text{v}/\text{kg}_\text{a}$ (see chart for the absolute humidity corresponding to the dew point on the saturation curve at 5 °C).

Drawing a horizontal line from that dew point to the right until the dry bulb temperature of 23 °C is reached, one obtains point P. This point is situated on the curve $\phi = 32$ %, representing the maximum relative humidity that can be accepted in the room without condensation on windows.

$$p = 748 \text{ mmHg} = 0.99708 \text{ bar}$$

$$\omega = 0.622 \frac{p_v}{p - p_v} \quad (\text{kg}_\text{v}/\text{kg}_\text{a})$$

$$p_v = \frac{\omega p}{0.622 + \omega} = \frac{0.0055 \times 0.99708}{0.622 + 0.0055} = 0.00874 \text{ bar}$$

$$p_a = p - p_v = 0.99708 - 0.00874 = 0.98834 \text{ bar}$$

The specific volume is $v \approx 0.845 \text{ m}^3/\text{kg}_\text{a}$. The density will be

$$\rho = \frac{1}{v} = \frac{1}{0.845} = 1.183 \text{ kg/m}^3$$

7.9. *Assumption*: Humid air is considered an ideal gas. All processes occur at constant atmospheric pressure (760 mmHg), so the standard psychrometric chart can be used.

The absolute humidity remains constant.

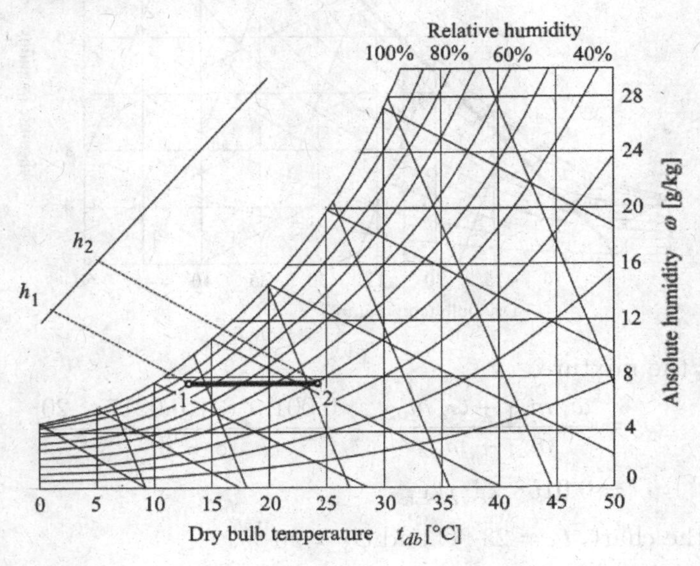

From the chart, for the initial state the specific enthalpy is $h_1 \approx 31.5$ kJ/kg$_a$. For the final state, $\phi_2 = 40$ %, $v_2 \approx 0.852$ m^3/kg$_a$ and $h_2 \approx 43$ kJ/kg$_a$. The corresponding density will be $\rho_2 = 1.174$ kg/m^3.

So, the final relative humidity is $\phi_2 = 40$ %.

The heat transfer rate is

$$\dot{Q} = \dot{m}\,\Delta h = \rho \dot{V}\,(h_2 - h_1) = 1.174 \times 65\,(43 - 31.5)$$

$$= 877.6 \text{ kJ/min} = 14.6 \text{ kW}$$

7.10. *Assumption*: Humid air is considered an ideal gas. All processes occur at constant atmospheric pressure (760 mmHg), so the standard psychrometric chart can be used.

From the psychrometric chart for $t_1 = 5$ °C and $\phi_1 = 20$ %:, $\omega_1 = 0.001$ kg$_v$/kg$_a$.

Also, for $t_2 = 30$ °C and $\phi_2 = 80$ %, $\omega_2 = 0.0217$ kg$_v$/kg$_a$.

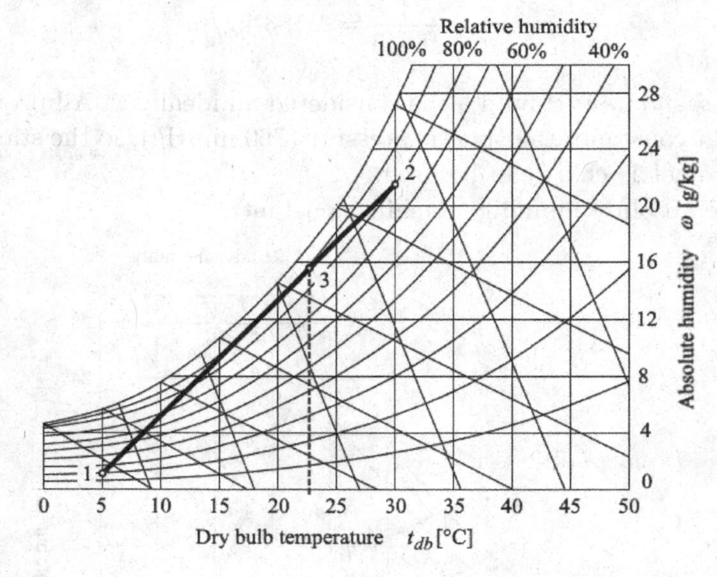

For the mixture,

$$\omega_3 = \frac{\omega_1\, m_{a1} + \omega_2\, m_{a2}}{m_{a1} + m_{a2}} = \frac{0.001 \times 8 + 0.0217 \times 20}{8 + 20}$$

$$= 0.0158 \text{ kg}_v/\text{kg}_a$$

From the chart, $t_3 = 23$ °C and $\phi_3 = 90$ %.

8.1. The problem can be solved using the $h - s$ (Mollier) diagram for steam.

Point 1: $p_1 = 60$ bar; $t_1 = 400$ °C $\Rightarrow h_1 \approx 3180$ kJ/kg.
Process 1–2: throttling; $h_2 = h_1$.
Point 2: $h_2 = 3180$ kJ/kg; $p_2 = 50$ bar; $t_2 \approx 390$ °C.
Process 2–3: adiabatic reversible (isentropic); $s_3 = s_2$.
Point 3: end of isentropic process; $p_3 = 2$ bar, $h_3 \approx 2510$ kJ/kg; $x_3 \approx 0.92$.
Process 2–3a: adiabatic (nonisentropic, therefore irreversible); $s_{3a} > s_3$.
Point 3a: end of adiabatic expansion; $p_{3a} = p_3 = 2$ bar.

$$h_{3a} = h_2 - \eta_T (h_2 - h_3)$$

$$h_{3a} = 3180 - 0.86(3180 - 2510) = 2604 \text{ kJ/kg}$$

The steam quality is $x_{3a} \approx 0.96$.
The power produced by the turbine is

$$P = \dot{m}\,(h_2 - h_{3a}) = \frac{50 \times 10^3}{3600}(3180 - 2604) = 8000 \text{ kW} = 8 \text{ MW}$$

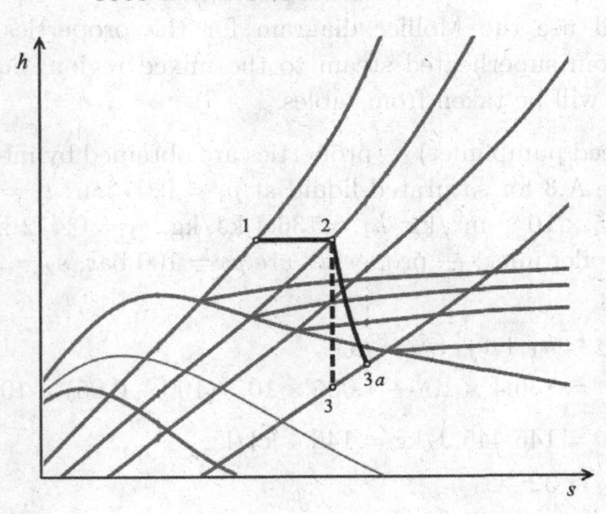

To summarize:

(a) temperature of steam at turbine inlet: $t_2 \approx 390$ °C
(b) steam quality at turbine exit: $x_{3a} \approx 0.96$ (irreversibility improves the steam quality)
(c) power produced: $P = 8$ MW.

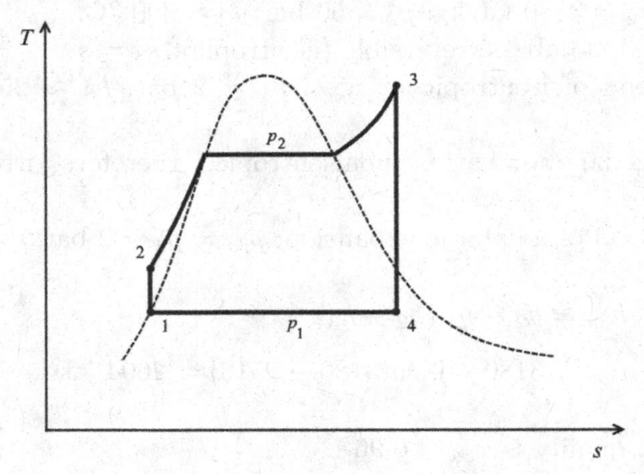

8.2. *Assumption*: The plant operates based on the ideal Rankine cycle: compression in the pump and expansion in the turbine are ideal (isentropic); heating in the boiler and condensation in the condenser are isobaric processes (no pressure losses).

We will use the Mollier diagram for the properties of steam ranging from superheated steam to the mixed region; liquid water properties will be taken from tables.

Point 1 (feed pump inlet) — properties are obtained by interpolation from Table A.3 for saturated liquid at $p_1 = 0.05$ bar: $t_1 = 32.56$ °C, $v_{f1} = 1.005 \times 10^{-3}$ m^3/kg; $h_1 = 136.4$ kJ/kg, $s_1 = 0.472$ kJ/(kg K).
Point 2 (boiler inlet) — properties are $p_2 = 100$ bar, $s_2 = s_1 = 0.472$ kJ/(kg K),

$$\begin{aligned} h_2 &\approx h_1 + v_{f1}\,(p_2 - p_1) \\ &= 136.4 \times 10^3 + 1.005 \times 10^{-3}(100 - 0.05) \times 10^5 \\ &= 146\,445 \text{ J/kg} = 146.4 \text{ kJ/kg} \end{aligned}$$

$$t_2 \approx 32 \text{ °C}$$

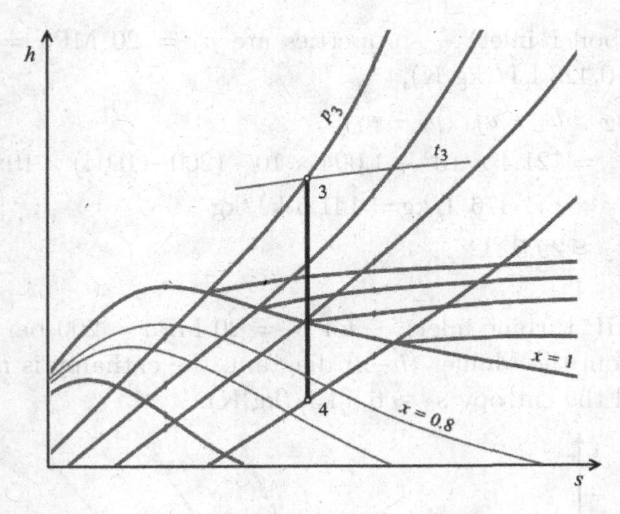

Point 3 (turbine inlet) — for $p_3 = 100$ bar and $t_3 = 500$ °C, from the Mollier ($h - s$) diagram, the enthalpy is $h_3 \approx 3375$ kJ/kg and the entropy $s_3 \approx 6.6$ kJ/(kg K).
Point 4 (condenser inlet) — from the Mollier diagram: $s_4 = s_3 \approx 6.6$ kJ/(kg K); $h_4 \approx 2000$ kJ/kg; $t_4 = t_1 = 32.56$ °C; $x \approx 0.78$.

It is to be noticed that the work consumed by the feed pump $w_{\text{pump.in}} \approx v_{f1}(p_1 - p_2) = -10$ kJ/kg is much smaller than the work produced by the steam turbine, $w_{\text{turb.out}} = h_3 - h_4 = 1375$ J/kg; the back work ratio r_{bw} is about 0.0073. In other words, $w_{\text{net}} \approx w_{\text{turb.out}}$. Therefore points 1 and 2 almost coincide on the $T{-}s$ diagram.

8.3. *Assumption*: The plant operates based on the ideal Rankine cycle: compression in the pump and expansion in the turbine are ideal (isentropic); heating/reheating in the boiler and condensation in the condenser are isobaric processes (no pressure losses).

We will use the Mollier diagram for the properties of steam ranging from superheated steam to the mixed region; liquid water properties will be taken from tables.

Point 1 (feed pump inlet) — properties are obtained from thermodynamic Table A.3 for saturated liquid at $p_1 = 0.004$ MPa $= 0.04$ bar: $t_1 = 28.96$ °C, $v_{f1} = 1.00 \times 10^{-3}$ m^3/kg; $h_1 = 121.4$ kJ/kg, $s_1 = 0.4224$ kJ/(kg K).

Point 2 (boiler inlet) — properties are $p_2 = 20$ MPa $= 200$ bar, $s_2 = s_1 = 0.423$ kJ/(kg K),

$$h_2 \approx h_1 + v_{f1}(p_2 - p_1)$$
$$= 121.4 \times 10^3 + 1.004 \times 10^{-3}(200 - 0.04) \times 10^5$$
$$= 141\ 476 \text{ J/kg} = 141.5 \text{ kJ/kg}$$
$$t_2 \approx 29.5\ ^\circ\text{C}$$

Point 3 (HP turbine inlet) — for $p_3 = 20$ MPa $= 200$ bar and $t_3 = 700$ °C, from the Mollier (h–s) diagram, the enthalpy is $h_3 \approx 3800$ kJ/kg and the entropy $s_3 \approx 6.8$ kJ/(kg K).

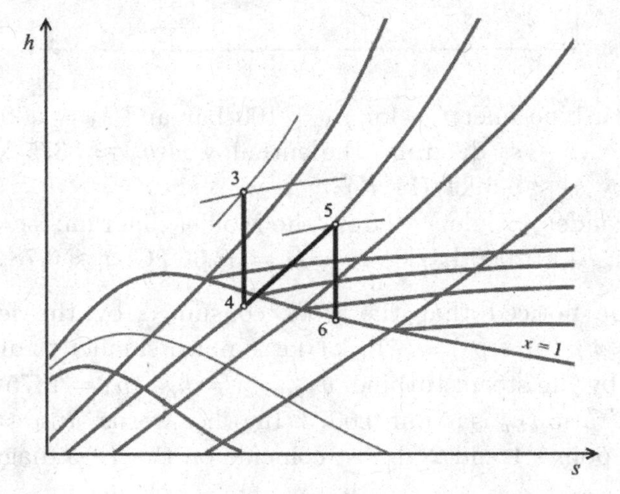

Point 4 (reheater inlet) — from the Mollier diagram: $s_4 = s_3 \approx 6.6$ kJ/(kg K); $h_4 \approx 2690$ kJ/kg; $x \approx 0.98$.
Point 5 (LP turbine inlet) — for $p_5 = p_4 = 0.4$ MPa $= 4$ bar and $t_5 = 600$ °C, from the Mollier (h–s) diagram, the enthalpy is $h_5 \approx 3700$ kJ/kg and the entropy $s_5 \approx 8.45$ kJ/(kg K).
Point 6 (condenser inlet) — from the Mollier diagram: $s_6 = s_5 \approx 8.45$ kJ/(kg K); $h_6 \approx 2550$ kJ/kg; $t_6 = t_1 = 28.96$ °C; $x \approx 1$.

(a)

$$\eta_{\text{th}} = \frac{w_{\text{net}}}{q_{\text{in}}} = \frac{w_{\text{turb.out}} - |w_{\text{pump.in}}|}{q_{in}}$$
$$= \frac{(h_3 - h_4) + (h_5 - h_6) - |h_1 - h_2|}{(h_3 - h_2) + (h_5 - h_4)}$$

$$\eta_{th} = \frac{(3800-2690)+(3700-2550)-|121.4-141.5|}{(3800-141.5)+(3700-2690)}$$
$$\approx 0.48 = 48~\%$$

Note: Without reheat, assuming the steam expands in the turbine from point 3 (200 bar, 700 °C) to point 4′ (0.04 bar), the enthalpy would be $h_{4'} \approx 2020$ kJ/kg and the steam quality $x \approx 0.78$. The thermal efficiency would have been

$$\eta_{th} = \frac{w_{\text{net}}}{q_{\text{in}}} = \frac{w_{\text{turb.out}} - |w_{\text{pump.in}}|}{q_{\text{in}}} = \frac{(h_3 - h_{4'}) - |h_1 - h_2|}{h_3 - h_2}$$
$$= \frac{(3800-2020)-|121.4-141.5|}{3800-141.5}$$
$$= 0.481 = 48.1~\%$$

The thermal efficiency is practically the same, but reheat improves substantially the final steam quality.

(b) The mass flow rate of steam is

$$\dot{m} = \frac{P}{w_{\text{net}}} = \frac{P}{(h_3-h_4)+(h_5-h_6)-|h_1-h_2|}$$
$$= \frac{50\times 10^3}{(3800-2690)+(3700-2550)-|121.4-141.5|}$$
$$\dot{m} = 22.3 \text{ kg/s} = 80.36 \text{ t/h}$$

8.4. *Assumption*: The plant operates based on the ideal Rankine cycle with superheat: compression in the pump and expansion in the turbine are ideal (isentropic); heating in the boiler and condensation in the condenser are isobaric processes (no pressure losses).

The Mollier diagram will be used to determine the properties of steam ranging from superheated steam to the mixed region; liquid water properties will be taken from tables.

The $T\!-\!s$ and $h\!-\!s$ diagrams are the same as the ones used for Problem 8.2.

Point 1: $p_1 = p_4 = 0.08$ bar; $t_1 = t_{f1} = 41.5$ °C $\Rightarrow h_1 = h_f = 173.8$ kJ/kg; $v_f = 1.008 \times 10^{-3}$ m^3/kg.

Point 2: $p_2 = 8$ MPa $= 80$ bar;

$$w_{\text{pump.in}} \cong v_f(p_1 - p_2) = 1.008 \times 10^{-3}(0.06 - 80)10^5 = -8.06 \text{ kJ/kg}$$
$$h_2 = h_1 - w_{\text{pump.in}} = 181.8 \text{ kJ/kg}$$

Point 3: $p_3 = 80$ bar; $t_3 = 500\ °\text{C} \Rightarrow h_3 = 3400$ kJ/kg; $s_3 = s_4 = 6.7$ kJ/(kg K).
Point 4: $p_4 = 0.06$ bar; $t_4 = t_1 = t_{\text{f1}} \Rightarrow h_4 = 2060$ kJ/kg; $x_4 = 0.79$

(a)

$$\begin{aligned} w_{34} &= h_3 - h_4 = 3400 - 2095 \\ &= 1305 \text{ kJ/kg} \quad (w_{34} \gg |w_{\text{pump.in}}|) \\ &\Rightarrow w_{\text{net.out}} \cong w_{34} \quad \text{Points 1 and 2 are almost identical!} \\ \dot{m}_{\text{steam}} &= \frac{P}{w_{34}} = \frac{80\,000}{1305} = 61.30 \text{ kg/s} \\ \dot{Q}_{23} &= \dot{m}_{\text{steam}}(h_3 - h_2) \cong \dot{m}_{steam}(h_3 - h_1) \\ &= 61.30(3400 - 173.8) = 197.8 \times 10^3 \text{ kW} \end{aligned}$$

(b)

$$\eta_{\text{th}} = \frac{P}{\dot{Q}_{23}} = \frac{80}{197.8} = 0.40 = 40\ \%$$

(c)

$$\dot{Q}_{\text{steam}} = \dot{Q}_{\text{cw}}, \quad \dot{m}_{\text{steam}}(h_4 - h_1) = \dot{m}_{\text{cw}}\, c_{\text{pw}}(t_{\text{cw.out}} - t_{\text{cw.in}})$$

The mass flow rate of the cooling water is

$$\dot{m}_{\text{cw}} = \frac{\dot{m}_{\text{steam}}(h_4 - h_1)}{c_{\text{pw}}(t_{\text{cw.out}} - t_{\text{cw.in}})} = \frac{61.3\,(2060 - 173.8)}{4.18\,(38 - 17)} = 1317 \text{ kg/s}$$

Cooling such a large quantity of water requires cooling towers.

For comparison, the ratio of the mass flow rates is $\frac{\dot{m}_{\text{cw}}}{\dot{m}_{\text{steam}}} = \frac{1317}{61.3} = 21.48$.

Note: For reference only; if the heating power of coal has an average value of 15 000 kJ/kg (lignite) and the thermal efficiency of the boiler is 90%, the coal consumption is

$$\dot{m}_{\text{coal}} = \frac{\dot{Q}_{23}}{\eta_b\, HV} = \frac{197\,800}{0.9 \times 15\,000}$$

$$= 14.65 \text{ kg/s} = 52.75 \text{ t/h} = 1266 \text{ t/day}$$

9.1. *Assumption*: Cold air-standard assumptions. The working fluid is air having $k = 1.4$.

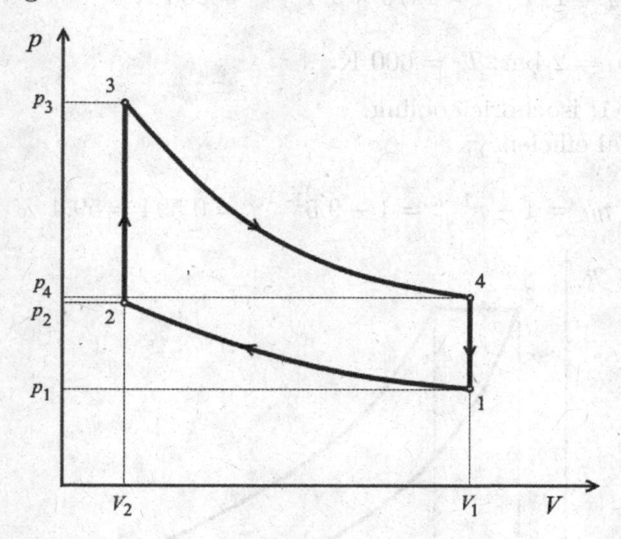

Point 1: $p_1 = 100$ kPa $= 1$ bar; $T_1 = 273 + 27 = 300$ K.

Process 1–2: adiabatic compression

$$p_2 = p_1 \left(\frac{V_1}{V_2}\right)^k = p_1 r^k = 100 \times 9.5^{1.4} = 2338 \text{ kPa} = 23.4 \text{ bar}$$

$$T_2 = T_1\, r^{k-1} = 300 \times 9.5^{1.4-1} = 738 \text{ K}$$

Point 2: $p_2 = 23.4$ bar; $T_2 = 738$ K.

Process 2–3: isochoric heating

$$p_3 = 2\,p_2 = 2 \times 2338 = 4676 \text{ kPa} = 46.8 \text{ bar}$$

$$T_3 = T_2\,\frac{p_3}{p_2} = 738\ \times 2 = 1476 \text{ K}$$

Point 3: $p_3 = 2p_2 = 46.8$ bar; $T_3 = 1476$ K.

Process 3–4: adiabatic expansion

$$p_4 = p_3 \left(\frac{V_3}{V_4}\right)^k = p_3\, r^{-k} = 2\, p_2 r^{-k}$$

$$= 2p_1 r^k r^{-k} = 2p_1 = 2 \times 100 = 200 \text{ kPa} = 2 \text{ bar}$$

$$T_4 = T_3\, r^{1-k} = 1476 \times 9.5^{1-1.4} = 600 \text{ K}$$

Point 4: $p_4 = 2$ bar; $T_4 = 600$ K.

Process 4–1: isochoric cooling.

Thermal efficiency:

$$\eta_{th} = 1 - r^{1-k} = 1 - 9.5^{1-1.4} = 0.594 = 59.4\ \%$$

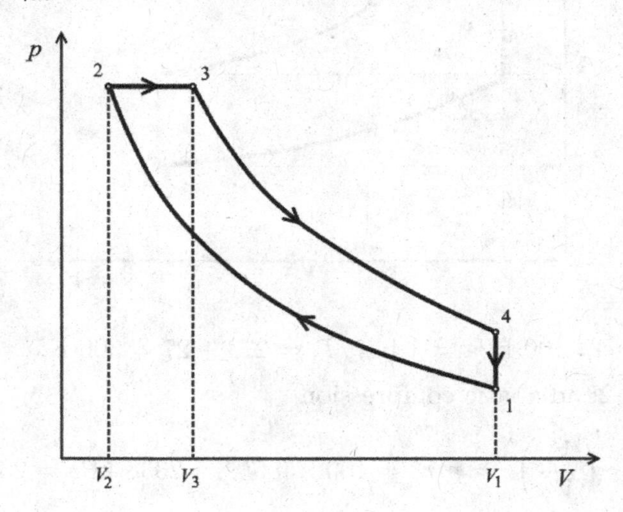

9.2. *Assumption*: cold air-standard assumptions. The working fluid is air having $k = 1.4$.

The thermal efficiency of the Diesel cycle can be calculated using the equation:

$$\eta_{\text{th,Diesel}} = 1 - \frac{1}{k}\left[\frac{r_c^k - 1}{r^{k-1} r_c - r^{k-1}}\right]$$

which requires the calculation of the cut-off ratio r_c:

$$r_c = \frac{T_3}{T_2}$$

T_3 is the maximum temperature of the cycle, therefore $T_3 = 3200$ K. Then,

$$T_2 = T_1 \left(\frac{v_2}{v_1}\right)^{k-1} = T_1\, r^{k-1} = (200 + 273) \times 20^{1.4-1} = 1567.7 \text{ K}$$

$$r_c = \frac{T_3}{T_2} = \frac{3200}{1567.7} = 2.04$$

$$\eta_{th} = 1 - \frac{1}{1.4}\,\frac{2.04^{1.4} - 1}{20^{1.4-1} \times 2.04 - 20^{1.4-1}} = 0.645 \approx 64.5 \text{ \%}$$

The mean effective pressure p_m is given by

$$p_m = \frac{w_{net}}{v_1 - v_2}$$

$$v_1 = \frac{RT_1}{p_1} = \frac{287 \times (200 + 273)}{200\ 000} = 0.6788 \text{ m}^3\text{/kg}$$

$$v_2 = \frac{v_1}{r} = \frac{0.6788}{20} = 0.0339\text{m}^3\text{/kg}$$

$$p_m = \frac{1200}{0.6788 - 0.0339} = 1861 \text{ kPa} = 18.6 \text{ bar}$$

9.3. *Assumption*: cold air-standard assumptions. The working fluid is air having $k = 1.4$.

$$r = \frac{v_1}{v_2} = 8, \quad r_p = \frac{p_3}{p_2} = \frac{T_3}{T_2}, \quad r_c = \frac{v_4}{v_3} = \frac{T_4}{T_3}$$

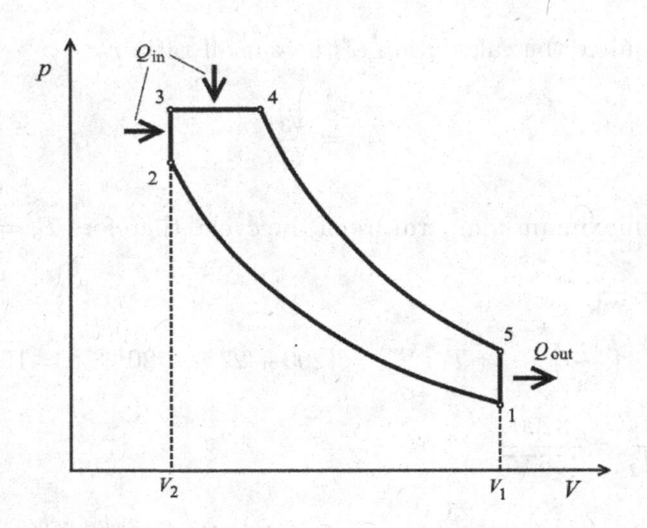

(a) $p_1 = 0.85$ bar $= 85\ 000$ Pa; $T_1 = 50 + 273 = 323$ K.

$$v_1 = \frac{RT_1}{p_1} = \frac{287 \times 323}{85\ 000} = 1.0906 \text{ m}^3/\text{kg}$$

Process 1–2: isentropic compression

$$v_2 = \frac{v_1}{r} = \frac{1.0906}{8} = 0.136 \text{ m}^3/\text{kg}$$
$$T_2 = T_1 r^{k-1} = 323 \times 8^{1.4-1} = 742 \text{ K}$$
$$p_2 = p_1 \left(\frac{v_1}{v_2}\right)^k = p_1 r^k = 85\ 000 \times 8^{1.4} = 1\ 562\ 230 \text{ Pa} = 15.62 \text{ bar}$$

Process 2–3: isochoric heating

$$v_3 = v_2 = 0.136 \text{ m}^3/\text{kg}$$
$$T_3 = T_2 \frac{p_3}{p_2} = T_2\, r_p = 742 \times 2 = 1484 \text{ K}$$
$$p_3 = r_p\, p_2 = 2 \times 1\ 562\ 230 = 3\ 124\ 460 \text{ Pa} = 31.24 \text{ bar}$$

Process 3–4: isobaric heating

$$v_4 = r_c\, v_3 = 1.2 \times 0.136 = 0.1632 \text{ m}^3/\text{kg}$$
$$T_4 = T_3 \frac{v_4}{v_3} = T_3\, r_c = 1484 \times 1.2 = 1781 \text{ K}$$
$$p_4 = p_3 = 31.24 \text{ Pa}$$

Process 4–5: isentropic expansion

$v_5 = v_1 = 1.0906 \text{ m}^3/\text{kg}$

$$T_5 = T_4 \left(\frac{v_4}{v_5}\right)^{k-1} = 1781 \times \left(\frac{0.1632}{1.0906}\right)^{1.4-1} = 833 \text{ K}$$

$$p_5 = p_4 \left(\frac{v_4}{v_5}\right)^{k} = 3124460 \times \left(\frac{0.1632}{1.0906}\right)^{1.4} = 219433 \text{ Pa} = 2.19 \text{ bar}$$

Process 5–1: isochoric cooling

(b)

$$q_{\text{in}} = q_{23} + q_{34} = c_v(T_3 - T_2) + c_p(T_4 - T_3)$$

For cold air, considering $c_p = 1004$ J/(kg K); $c_v = 717$ J/(kg K)

$$q_{\text{in}} = 717 \times (1484 - 742) + 1004 \times (1781 - 1484)$$
$$= 830\,067 \text{ J/kg} = 830 \text{ kJ/kg}$$

(c) $w_{\text{net}} = q_{\text{in}} - |q_{\text{out}}|$

$$|q_{\text{out}}| = c_v\,(T_5 - T_1) = 717 \times (833 - 323)$$
$$= 366\,277 \text{ J/kg} = 366 \text{ kJ/kg}$$

$$w_{\text{net}} = 830 - 366 = 464 \text{ kJ/kg}$$

(d) The thermal efficiency of the cycle is given by

$$\eta_{\text{th}} = \frac{w_{\text{net}}}{q_{\text{in}}} = \frac{464}{830} = 0.559 \approx 56 \text{ \%}$$

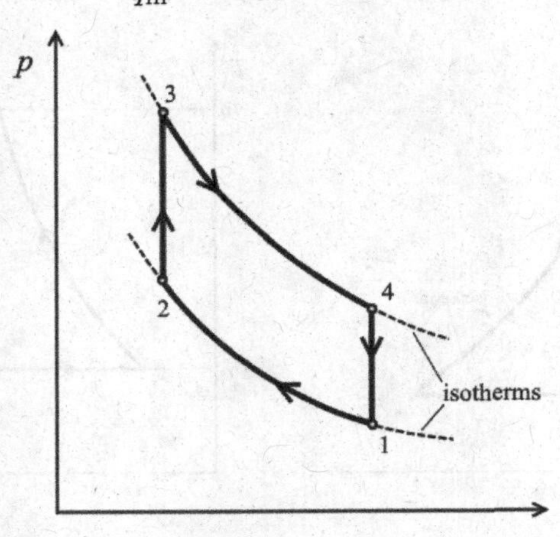

9.4. *Assumption*: cold air-standard assumptions. The working fluid is air, $R = 287$ J/(kg K).

$$p_2 = p_1 \frac{v_1}{v_2} = p_1 r = 100\ 000 \times 6 = 600\ 000 \text{ Pa} = 6 \text{ bar}$$

$$p_{\max} = p_3 = p_2 \frac{T_3}{T_2} = 6 \times \frac{1100 + 273}{25 + 273} = 27.6 \text{ bar}$$

$$\eta_{\text{th}} = 1 - \frac{T_C}{T_H} = 1 - \frac{298}{1373} = 0.783 = 78.3\ \%$$

$$w_{\text{out}} = w_{34} + w_{12} = R\left(T_3 \ln \frac{v_4}{v_3} + T_1 \ln \frac{v_2}{v_1}\right)$$

$$= R\left(T_3 \ln r + T_1 \ln \frac{1}{r}\right)$$

$$w_{\text{out}} = 287\left(1373 \ln 6 + 298 \ln \frac{1}{6}\right) = 552\ 803 \text{ J/kg} = 552.8 \text{ kJ/kg}$$

$$q_{\text{in}} = \frac{w_{\text{out}}}{\eta_{\text{th}}} = \frac{552.8}{0.783} = 706.0 \text{ kJ/kg}$$

9.5. *Assumption*: cold air-standard assumptions. The working fluid is air, $R = 287$ J/(kg K).

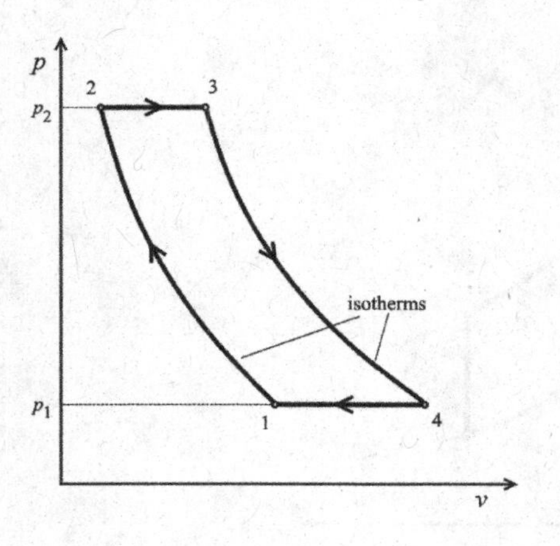

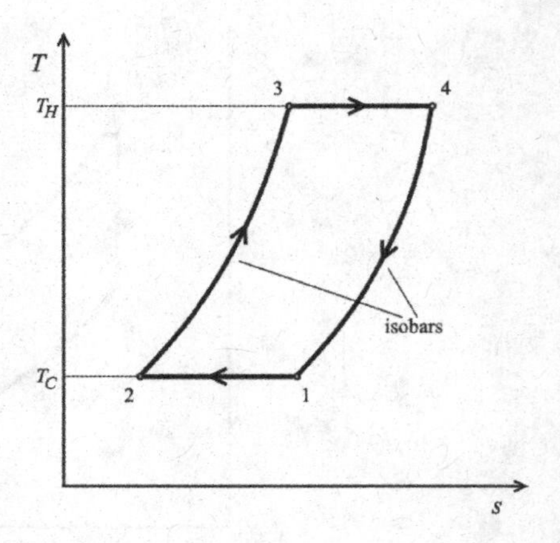

Known values:

$$T_1 = T_2 = 100 + 273 = 373 \text{ K}$$
$$p_1 = p_4 = 100 \text{ kPa} = 100\,000 \text{ Pa}$$
$$T_3 = T_4 = 600 + 273 = 873 \text{ K}$$
$$p_2 = p_3 = 1000 \text{ kPa} = 1\,000\,000 \text{ Pa}$$

The power output: $P = \dot{m}\, w_{\text{out}}$

$$w_{\text{out}} = w_{12} + w_{23} + w_{34} + w_{41}$$
$$w_{\text{out}} = RT_1 \ln \frac{v_2}{v_1} + p_2(v_3 - v_2) + RT_3 \ln \frac{v_3}{v_4} + p_1(v_1 - v_4)$$

The volumes in the four characteristic points of the circle need to be determined first:

$$v_1 = \frac{RT_1}{p_1} = \frac{287 \times 373}{100\,000} = 1.07 \text{ m}^3/\text{kg}$$
$$v_2 = \frac{RT_2}{p_2} = \frac{287 \times 373}{1\,000\,000} = 0.107 \text{ m}^3/\text{kg}$$
$$v_3 = \frac{RT_3}{p_3} = \frac{287 \times 873}{1\,000\,000} = 0.250 \text{ m}^3/\text{kg}$$
$$v_4 = \frac{RT_4}{p_4} = \frac{287 \times 873}{100\,000} = 2.50 \text{ m}^3/\text{kg}$$
$$w_{\text{out}} = 287 \times 373 \times \ln \frac{0.107}{1.07} + 10^6(0.250 - 0.107)$$
$$+ 287 \times 873 \times \ln \frac{0.250}{2.50} + 10^5(1.07 - 2.50)$$
$$w_{\text{out}} = 330\,421 \text{ J/kg} = 330.4 \text{ kJ/kg}$$
$$P = 1 \times 330.4 = 330.4 \text{ kW}$$

9.6. *Assumption*: cold air-standard assumptions. The working fluid is air, $R = 287$ J/(kg K).

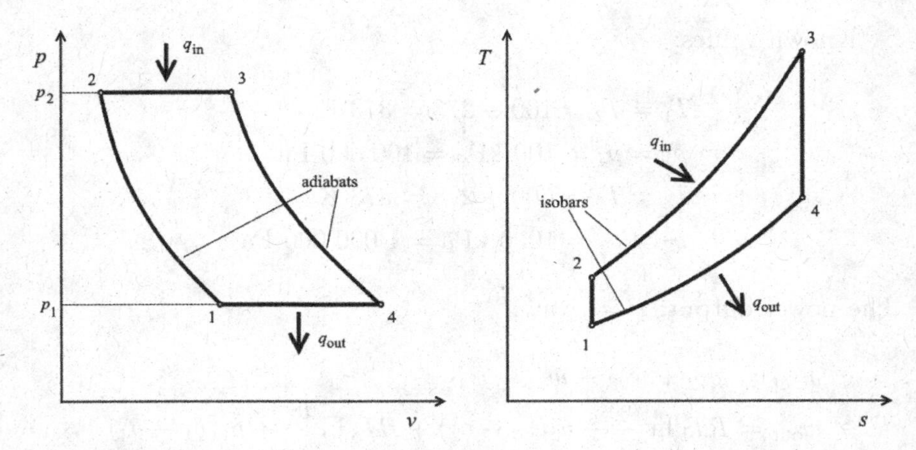

Given values:

$$T_1 = 27 + 273 = 300 \text{ K}$$

$$p_1 = 100 \text{ kPa} = 100\ 000 \text{ Pa}$$

$$T_3 = T_{\max} = 800 + 273 = 1073 \text{ K}$$

$$r_p = p_2/p_1 = p_3/p_4 = 6.5$$

The compressor work $w_C = h_1 - h_2 = c_p(T_1 - T_2)$

The turbine work $w_T = h_3 - h_4 = c_p(T_3 - T_4)$

$$T_2 = T_1 \left(\frac{p_2}{p_1}\right)^{\frac{k-1}{k}} = T_1\, r_p^{(k-1)/k} = 300 \times 6.5^{(1.4-1)/1.4} = 512.1 \text{ K}$$

$$T_4 = T_3 \left(\frac{p_4}{p_3}\right)^{\frac{k-1}{k}} = T_3 \left(\frac{1}{r_p}\right)^{(k-1)/k}$$

$$= 1073 \times \left(\frac{1}{6.5}\right)^{(1.4-4)/1.4} = 628.5 \text{ K}$$

$$r_{\text{bw}} = \frac{|w_C|}{w_T} = \frac{c_p(T_2 - T_1)}{c_p(T_3 - T_4)} = \frac{T_2 - T_1}{T_3 - T_4}$$

$$= \frac{512.1 - 300}{1073 - 628.5} = 0.477 = 47.7\ \%$$

$$\eta_{\text{th}} = 1 - r_p^{(1-4)/k} = 1 - 6.5^{(1-1.4)/1.4} = 0.414 = 41.4\ \%$$

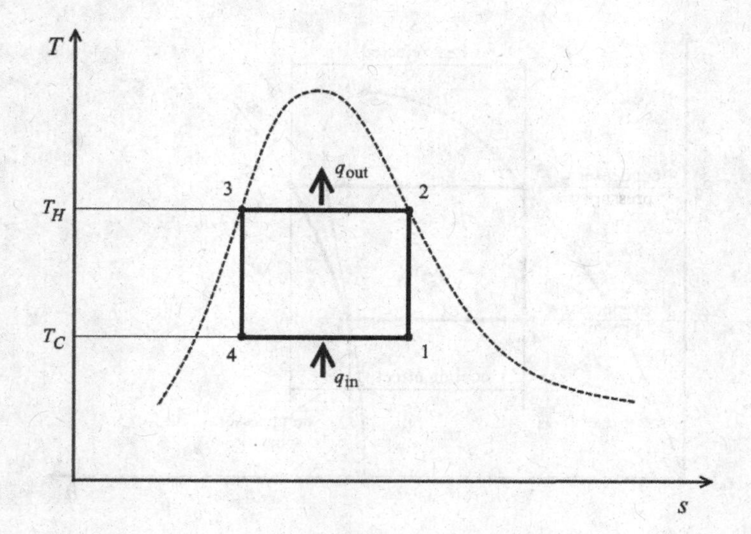

10.1. *Assumption*: The system operates under steady conditions. Kinetic and potential energy changes are negligible. All processes are internally reversible.

Known values:

$$T_H = 30 + 273 = 303 \text{ K}$$

$$T_C = -15 + 273 = 258 \text{ K}$$

$$\beta_C = \frac{T_C}{T_H - T_C} = \frac{258}{303 - 258} = 5.73$$

$$q_{\text{out}} = h_3 - h_2$$

From the refrigerant tables, at saturation,

$$h_3 = h_{f@30^\circ\text{C}} = 241.72 \text{ kJ/kg}$$

$$h_2 = h_{g@30^\circ\text{C}} = 414.82 \text{ kJ/kg}$$

$$q_{\text{out}} = -h_{fg} = 241.8 - 415.1 = -173.1 \text{ kJ/kg}$$

$$q_{\text{in}} = |q_{\text{out}}| \frac{T_C}{T_H} = 173.1 \times \frac{258}{303} = 147.4 \text{ kJ/kg}$$

$$w_{\text{net.in}} = q_{\text{in}} - |q_{\text{out}}| = 147.4 - 173.1 = -25.7 \text{ kJ/kg}$$

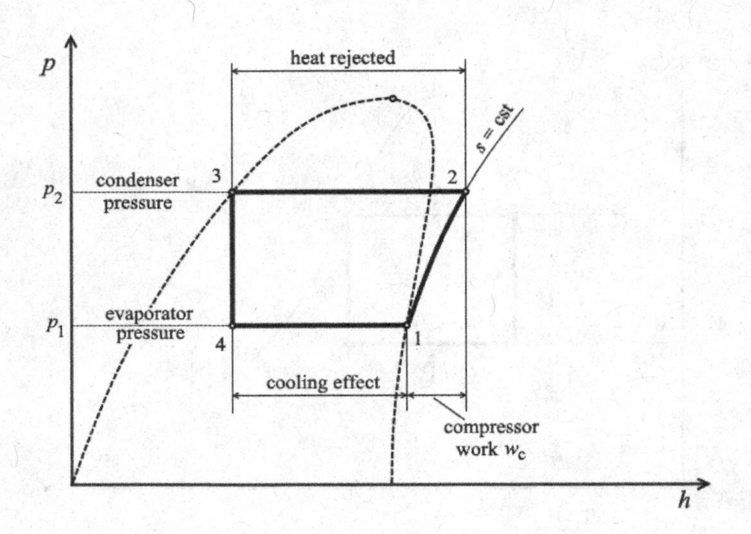

10.2. *Assumption*: The system operates under steady conditions. Kinetic and potential energy changes are negligible.

For a reasonable size of the heat exchangers, an average temperature difference of 10 °C is recommended between fluids. Therefore

$$t_1 = t_4 = 5 - 10 = -5\ °\text{C}; \quad t_3 = 24 + 10 = 34\ °\text{C}.$$

Saturation pressures corresponding to these temperatures can be found in Table A.5: $p_1 = p_4 = p_{\text{evap}} = 243.34$ kPa ≈ 2.4 bar;

$$p_2 = p_3 = p_{\text{cond}} = 862.63 \text{ kPa} \approx 8.6 \text{ bar}.$$

These are, therefore, the recommended evaporator and condenser pressures.

10.3. *Assumption*: The system operates under steady conditions. Kinetic and potential energy changes are negligible. Refer to the p–h diagram for Problem 10.2.

From the saturation tables for R134a: $h_1 = h_{g@320\text{ kPa}} = 400.0$ kJ/kg; $h_3 = 275.4$ kJ/kg.

From the p–h diagram for R134a: $h_2 \approx 430$ kJ/kg.

The power input is $P = \dot{m}\,w_C = \dot{m}(h_2 - h_1) = 0.1 \times (430 - 400.0) = 3.0$ kW.

The coefficient of performance is

$$\gamma = \frac{h_2 - h_3}{h_2 - h_1} = \frac{430 - 275.4}{430 - 400.0} = 5.15$$

10.4. *Assumption*: The system operates under steady conditions. Kinetic and potential energy changes are negligible. Refer to the p–h diagram for Problem 10.2.

From the saturation tables for R134a: $h_1 = h_{g@130\ \mathrm{kPa}} = 386.3$ kJ/kg; $h_3 = 240.5$ kJ/kg.

From the p–h diagram for R134a: $h_2 \approx 422$ kJ/kg.

$$h_4 = h_3 = 240.4 \text{ kJ/kg}$$

The heat removed from the refrigerated space (the evaporator):

$$\dot{Q}_{\mathrm{in}} = \dot{m}(h_1 - h_4) = 0.06 \times (386.3 - 240.4) = 8.75 \text{ kW}$$

The compressor power input is

$$P = |\dot{W}_C| = \dot{m}|w_C| = \dot{m}|h_1 - h_2| = 0.06 \times |386.3 - 422| = 2.14 \text{ kW}$$

The heat rejected to the environment in the condenser:

$$\dot{Q}_{\mathrm{out}} = \dot{m}(h_3 - h_2) = 0.06 \times (240.4 - 422) = -10.90 \text{ kW}$$

The coefficient of performance:

$$\gamma = \frac{\dot{Q}_{\mathrm{out}}}{\dot{W}_C} = \frac{-10.90}{-2.14} = 5.09$$

11.1. *Assumption*: The air is considered an ideal gas ($k = 1.4$). The ideal gas equation of state can be applied. $R = 287$ J/(kg K). The flow process is isentropic and steady.

Known values: $T_1 = 150 + 273 = 423$ K; $p_1 = 200\,000$ Pa; $p_2 = 100\,000$ Pa.

The critical pressure ratio for air is $(p_2/p_1)_{\mathrm{cr}} = 0.528$. For the case considered, $p_2/p_1 = 100/200 = 0.5 < (p_2/p_1)_{\mathrm{cr}}$. Therefore, for the fluid to be continuously accelerated, the nozzle has to be convergent divergent.

$$v_1 = \frac{RT_1}{p_1} = \frac{287 \times (150 + 273)}{200\,000} = 0.607 \text{ m}^3/\text{kg}$$

$$V_2 = \sqrt{\frac{2k}{1-k} p_1 v_1 \left[\left(\frac{p_2}{p_1} \right)^{(k-1)/k} - 1 \right]}$$

$$= \sqrt{\frac{2 \times 1.4}{1 - 1.4} \times 200\,000 \times 0.607 \times \left[\left(\frac{100\,000}{200\,000} \right)^{\frac{1.4-1}{1.4}} - 1 \right]}$$

$$V_2 = 390.7 \text{m/s}$$

The exit temperature is

$$T_2 = T_1 \left(\frac{p_2}{p_1} \right)^{\frac{k-1}{k}} = 423 \times \left(\frac{100\,000}{200\,000} \right)^{\frac{1.4-4}{1.4}} = 347 \text{ K}$$

The mass flow rate of air is

$$\dot{m} = \frac{A_2 V_2}{v_2} = \frac{A_2 V_2 p_2}{RT_2} = \frac{0.028 \times 390.7 \times 100\,000}{287 \times 347} \approx 11.0 \text{ kg/s}$$

11.2. *Assumption*: The air is considered an ideal gas ($k = 1.4$). The ideal gas equation of state can be applied. $R = 287$ J/(kg K). The flow process is isentropic and steady.

The local speed of sound is

$$c = \sqrt{kRT} = \sqrt{1.4 \times 287 \times 423} = 412.3 \text{ m/s}$$

The local Mach number is

$$Ma = \frac{V}{c} = \frac{460}{412.3} = 1.12 > 1$$

The flow is, therefore, supersonic.

11.3. *Assumption*: The air is considered an ideal gas ($k = 1.4$). The ideal gas equation of state can be applied. $R = 287$ J/(kg K). The flow process is isentropic and steady.

$$p_1 = 200 \text{ kPa} = 200\,000 \text{ Pa}; \quad T_1 = 350 \text{ K}.$$
$$c = \sqrt{k\,RT_1} = \sqrt{1.4 \times 287 \times 350} = 375 \text{ m/s}$$
$$V_2 = Ma \times c = 1.2 \times 375 = 450 \text{ m/s}$$

The flow being isentropic, $\frac{T_2}{T_1} = (\frac{p_2}{p_1})^{\frac{k-1}{k}}$. Also

$$V_2 = \sqrt{\frac{2k}{1-k} p_1 v_1 \left[\left(\frac{p_2}{p_1}\right)^{(k-1)/k} - 1\right]}$$
$$= \sqrt{\frac{2k}{1-k} RT_1 \left(\frac{T_2}{T_1} - 1\right)} = \sqrt{\frac{2k}{1-k} R\,(T_2 - T_1)}$$
$$T_2 = T_1 + \frac{V_2^2(1-k)}{2kR} = 350 + \frac{450^2 \times (1-1.4)}{2 \times 1.4 \times 287} \approx 249 \text{ K}$$

11.4. *Assumption*: Nitrogen is considered an ideal gas ($k = 1.4$). The ideal gas equation of state can be applied. $R = 296.8$ J/(kg K). The flow process is isentropic and steady. Nitrogen does not change its phase during expansion. The pressure inside the vessel is not significantly affected by the leak. The outside pressure is considered $p_2 = 100$ kPa.

The pressure ratio is $\frac{p_2}{p_1} = \frac{100}{250} = 0.4 < (\frac{p_2}{p_1})_{\text{cr}}$. Therefore the hole will behave as a converging nozzle and the flow will be choked at the critical parameters. The mass flow rate is

$$\dot{m} = A_2 \sqrt{k \left(\frac{2}{k+1}\right)^{(k+1)/(k-1)} \frac{p_1}{v_1}}$$
$$A_2 = \pi \frac{d^2}{4} = \pi \times \frac{0.0001^2}{4} = 7.85 \times 10^{-9} \text{ m}^2$$
$$v_1 = \frac{RT_1}{p_1} = 0.3478 \text{ m}^3/\text{kg}$$
$$\dot{m} = 2.883 \times 10^{-6} \text{ kg/s} = 0.0104 \text{ kg/h}$$

The actual leak rate is smaller, due to the fact that a hole with sharp edges is not exactly a convergent nozzle:

$$\dot{m}_{\text{leak}} = c_d\,\dot{m} = 0.67 * 0.0104 = 0.00695 \text{ kg/h} = 6.95 \text{ g/h}$$

12.1. (c)

12.2. (c)

12.3. (a)

12.4. (a)

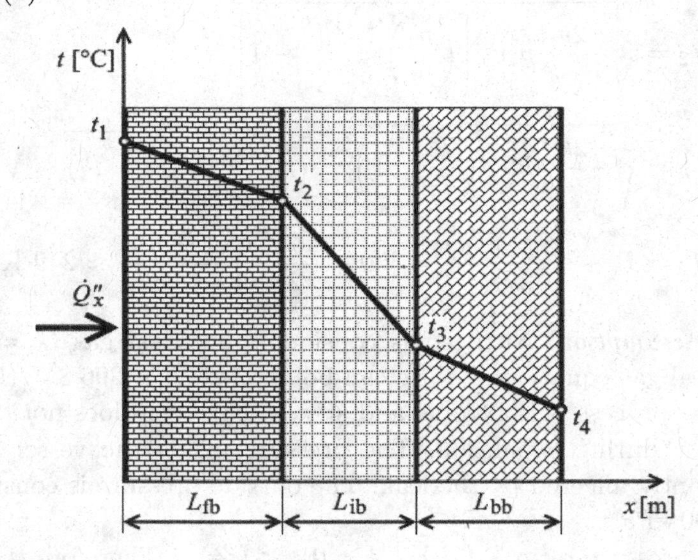

13.1. *Assumption*: Steady operating conditions exist. Heat transfer is one-dimensional. Thermal conductivities are constant. The thermal contact resistance at interfaces is negligible.

Given:

$$L_{\text{fb}} = 23 \text{ cm} = 0.23 \text{ m}; \quad k_{\text{fb}} = 1 \text{ W/(mK)}$$
$$L_{\text{ib}} = 15 \text{ cm} = 0.15 \text{ m}; \quad k_{\text{ib}} = 0.15 \text{ W/(mK)}$$
$$L_{\text{bb}} = 20 \text{ cm} = 0.20 \text{ m}; \quad k_{\text{bb}} = 0.6 \text{ W/(mK)}$$
$$t_1 = 850\ °\text{C}; \quad t_4 = 50\ °\text{C}$$
$$L_{\text{air}} = 7 \text{ mm} = 0.007 \text{ m}; \quad k_{\text{air}} = 0.06 \text{ W/(mK)}.$$

(a)

$$\dot{Q}''_x = \frac{t_1 - t_4}{\frac{L_{fb}}{k_{fb}} + \frac{L_{ib}}{k_{ib}} + \frac{L_{bb}}{k_{bb}}} = \frac{850 - 50}{\frac{0.23}{1} + \frac{0.15}{0.15} + \frac{0.20}{0.6}} = 511.7 \text{ W/m}^2$$

The heat flux being constant throughout the thickness of the wall, one can write:

$$\dot{Q}''_x = \frac{t_1 - t_4}{\frac{L_{\text{fb}}}{k_{\text{fb}}} + \frac{L_{\text{ib}}}{k_{\text{ib}}} + \frac{L_{\text{bb}}}{k_{\text{bb}}}} = \frac{t_1 - t_2}{\frac{L_{\text{fb}}}{k_{\text{fb}}}} = \frac{t_2 - t_3}{\frac{L_{\text{ib}}}{k_{\text{ib}}}} = \frac{t_3 - t_4}{\frac{L_{\text{bb}}}{k_{\text{bb}}}}$$

Therefore

$$t_2 = t_1 - \dot{Q}''_x \frac{L_{\text{fb}}}{k_{\text{fb}}} = 850 - 511.7 \times \frac{0.23}{1} = 732.3 \text{ °C}$$

$$t_3 = t_4 + \dot{Q}''_x \frac{L_{\text{bb}}}{k_{\text{bb}}} = 50 + 511.7 \times \frac{0.20}{0.6} = 220.6 \text{ °C}$$

Temperature t_3 was calculated this way to avoid perpetuating a possible error in the calculation of t_2.

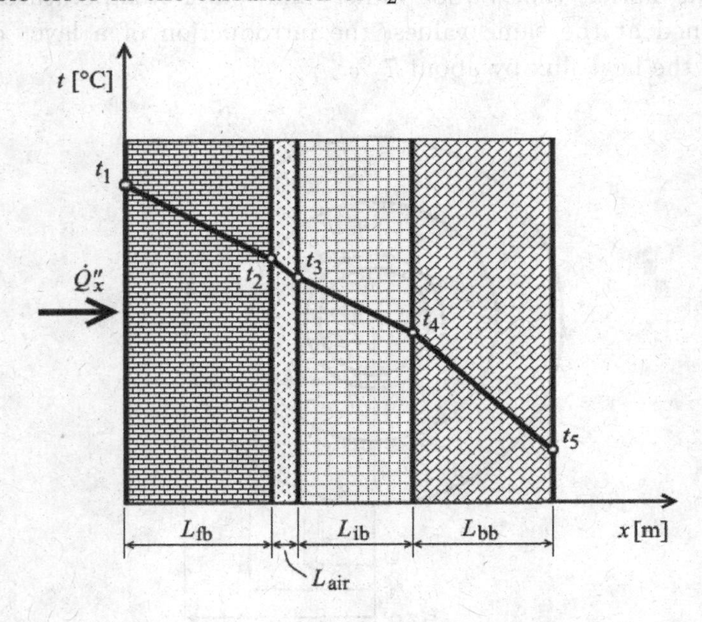

(b) When an intermediate layer of air is introduced, temperatures are t_1–t_5, with: $t_1 = 850$ °C and $t_5 = 50$ °C.

$$\dot{Q}''_x = \frac{t_1 - t_5}{\frac{L_{fb}}{k_{fb}} + \frac{L_{air}}{k_{air}} + \frac{L_{ib}}{k_{ib}} + \frac{L_{bb}}{k_{bb}}}$$

$$= \frac{t_1 - t_2}{\frac{L_{fb}}{k_{fb}}} = \frac{t_2 - t_3}{\frac{L_{air}}{k_{air}}} = \frac{t_3 - t_4}{\frac{L_{ib}}{k_{ib}}} = \frac{t_4 - t_5}{\frac{L_{bb}}{k_{bb}}}$$

$$\dot{Q}''_x = \frac{t_1 - t_5}{\frac{L_{fb}}{k_{fb}} + \frac{L_{air}}{k_{air}} + \frac{L_{ib}}{k_{ib}} + \frac{L_{bb}}{k_{bb}}}$$

$$= \frac{850 - 50}{\frac{0.23}{1} + \frac{0.007}{0.06} + \frac{0.15}{0.15} + \frac{0.20}{0.6}} = 476.2 \text{ W/m}^2$$

$$t_2 = t_1 - \dot{Q}''_x \frac{L_{fb}}{k_{fb}} = 850 - 476.2 \times \frac{0.23}{1} = 740.5 \text{ °C}$$

$$t_3 = t_2 - \dot{Q}''_x \frac{L_{air}}{k_{air}} = 850 - 476.2 \times \frac{0.007}{0.06} = 684.9 \text{ °C}$$

$$t_4 = t_{54} + \dot{Q}''_x \frac{L_{bb}}{k_{bb}} = 50 + 476.2 \times \frac{0.20}{0.6} = 208.7 \text{ °C}$$

One can notice that, when inner and outer temperatures are maintained at the same values, the introduction of a layer of air reduces the heat flux by about 7 %.

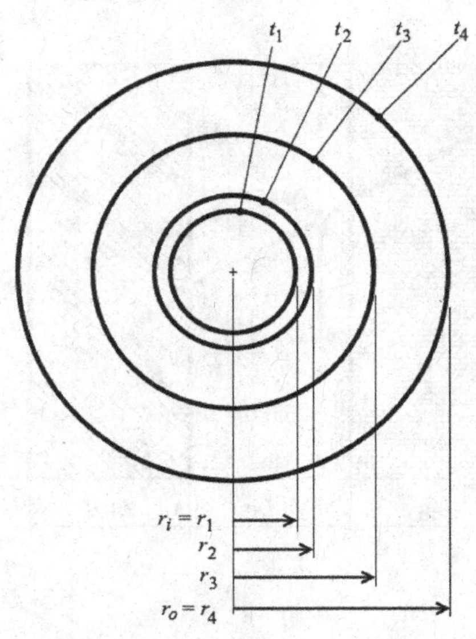

13.2. *Assumption*: Heat transfer is steady and one-dimensional (there is thermal symmetry about the center line and no temperature variation in the axial direction). Thermal conductivities are constant. The thermal contact resistance at interfaces is negligible.

Given:

$$L = 1\ \text{m}$$

$$d_i = 134.5\ \text{mm}; \quad r_i = r_1 = 67.25\ \text{mm}; \delta = 3.4\ \text{mm}$$

$$r_2 = r_1 + \delta = 70.65\ \text{mm}; \quad k = 64\ \text{W/(m K)}$$

$$\delta_1 = 30\ \text{mm}; \quad r_3 = r_2 + \delta_1 = 100.65\ \text{mm}; \quad k_1 = 0.037\ \text{W/(m K)}$$

$$\delta_2 = 50\ \text{mm}; \quad r_4 = r_3 + \delta_2 = 150.65\ \text{mm}; \quad k_2 = 0.14\ \text{W/(m K)}$$

$$t_i = t_1 = 280\ °\text{C}; \quad t_o = t_4 = 40\ °\text{C}$$

The heat flow rate in radial direction per unit length is

$$\dot{Q}'_r = \frac{\dot{Q}_r}{L} = \frac{2\pi(t_i - t_o)}{\frac{1}{k}\ln\frac{r_2}{r_1} + \frac{1}{k_1}\ln\frac{r_3}{r_2} + \frac{1}{k_2}\ln\frac{r_4}{r_3}}$$

$$= \frac{2\pi \times (280 - 40)}{\frac{1}{64} \times \ln\frac{70.65}{67.25} + \frac{1}{0.037} \times \ln\frac{100.65}{70.65} + \frac{1}{0.14} \times \ln\frac{150.65}{100.65}}$$

$$\dot{Q}'_r = 121.2\ \text{W/m}$$

When the layers of insulation are placed in reversed order,

$$\delta_1 = 50\ \text{mm}; \quad r_3 = r_2 + \delta_1 = 120.65\ \text{mm}; \quad k_1 = 0.14\ \text{W/(mK)}$$

$$\delta_2 = 30\ \text{mm}; \quad r_4 = r_3 + \delta_2 = 150.65\ \text{mm}; \quad k_2 = 0.037\ \text{W/(mK)}.$$

$$\dot{Q}'_r = \frac{\dot{Q}_r}{L} = \frac{2\pi(t_i - t_o)}{\frac{1}{k}\ln\frac{r_2}{r_1} + \frac{1}{k_1}\ln\frac{r_3}{r_2} + \frac{1}{k_2}\ln\frac{r_4}{r_3}}$$

$$= \frac{2\pi \times (280 - 40)}{\frac{1}{64} \times \ln\frac{70.65}{67.25} + \frac{1}{0.14} \times \ln\frac{120.65}{70.65} + \frac{1}{0.037} \times \ln\frac{150.65}{120.65}}$$

$$\dot{Q}'_r = 153.5\ \text{W/m}$$

This shows that the order of placement of insulation matters. The most efficient layer of insulation should be placed closer to the pipe.

13.3. *Assumption*: Steady operating conditions exist. The temperature of the surface is constant and uniform. The heat transfer coefficient is constant and uniform over the entire surface. The thermal properties are constant.

This is a case of free (natural) convection. Physical properties are considered at the film temperature t_f:

$$t_f = \frac{t_\infty + t_s}{2}$$

where $t_\infty = 20$ °C and $t_s = 140$ °C. Therefore $t_f = 80$ °C. At this temperature, the air properties are: $\beta = 0.00283$ 1/K; $\nu = 20.97 \times 10^{-6}$ m²/s; $Pr = 0.707$; $k = 0.0299$ W/(m K). The Grashof number is

$$\begin{aligned} Gr &= \frac{g\,\beta\,(t_s - t_\infty)L^3}{\nu^2} \\ &= \frac{9.81 \times 0.00283 \times (140 - 20) \times 1}{(20.97 \times 10^{-6})^2} = 3.791 \times 10^9 \end{aligned}$$

$$Gr\ Pr = 3.791 \times 10^9 \times 0.707 = 2.681 \times 10^9$$

Since $Gr\ Pr > 10^9$, $C = 0.13$; $n = 1/3$. The Nusselt number is

$$Nu = C(Gr\,Pr)^n = 0.13 \times (2.681 \times 10^9)^{1/3} = 180.574$$

The heat transfer coefficient can be obtained from the Nusselt number:

$$h = \frac{k\,Nu}{L} = \frac{0.0299 \times 180.574}{1} = 5.399 \text{ W/(m}^2\text{K)}$$

The heat transfer rate is

$$\dot{Q} = hA_s(t_s - t_\infty) = 5.399 \times 3.5 \times 1 \times (140 - 20) = 2268 \text{ W} = 2.268 \text{ kW}$$

13.4. *Assumption*: Steady-state conditions; one-dimensional incompressible viscous flow; fluid specific heats do not vary significantly with temperature; convection heat transfer coefficient is relatively constant along the tube.

Given: $t_m = 90$ °C; $t_s = 110$ °C; $d = 21$ mm $= 0.021$ m; $L = 2.5$ m; $V = 0.25$ m/s.

At t_m, the properties of water are: $\rho = 965.28$ kg/m^3; $\mu = 315 \times 10^{-6}$ (N s)/m^2; $k = 0.685$ W/(m K); $Pr = 1.97$.

The heat transfer rate is

$$\dot{Q} = hA_s(t_s - t_m) = h\,\pi dL(t_s - t_m)$$

The Reynolds number is calculated as

$$Re = \frac{\rho\,Vd}{\mu} = \frac{965.28 \times 0.25 \times 0.021}{315 \times 10^{-6}} = 16088$$

This is obviously a turbulent flow, since $Re > 2320$. Dittus-Boelter correlation can be used. For the heating of fluid,

$$Nu_d = 0.024\,Re_d^{0.8}\,Pr^{0.4} = 0.024 \times 16088^{0.8} \times 1.97^{0.4} = 72.98$$

$$h = \frac{k\,Nu}{d} = \frac{0.685 \times 72.98}{0.021} = 2380\,\mathrm{W/(m^2K)}$$

Now the heat transfer rate can be calculated:

$$\dot{Q} = h\,\pi dL(t_s - t_\infty) = 2380 \times \pi \times 0.021 \times 2.5 \times (110 - 90) = 7852\ \mathrm{W}$$

13.5. *Assumption*: Steady-state conditions; one-dimensional incompressible viscous flow; fluid specific heats do not vary significantly with temperature; convection heat transfer coefficient is constant along the tube.

Given: $t_\infty = 15$ °C; $t_s = 35$ °C; $d = 6$ mm $= 0.006$ m; $L = 2$ m; $V = 1.2$ m/s.

At the film temperature $t_f = (35 + 15)/2 = 25$ °C, the properties of water are: $\rho = 997.0$ kg/m^3; $\mu = 890 \times 10^{-6}$ (N s)/m^2; $k = 0.610$ W/(m K); $Pr = 6.10$.

The heat transfer rate is

$$\dot{Q} = hA_s(t_s - t_\infty) = h\pi dL(t_s - t_\infty)$$

The Reynolds number is calculated as

$$Re = \frac{\rho\,V\,d}{\mu} = \frac{997.0 \times 1.2 \times 0.006}{890 \times 10^{-6}} = 8066$$

Since $Re < 10^5$, Sparrow's correlation can be used.

$$Nu_D = 0.25 + (0.4Re_D^{1/2} + 0.06Re_D^{2/3})Pr^{0.37}$$

$$Nu_D = 0.25 + (0.4 \times 8066^{1/2} + 0.06 \times 8066^{2/3}) \times 6.10^{0.37} = 117.52$$

$$h = \frac{k\,Nu}{d} = \frac{0.610 \times 117.52}{0.006} = 11947 \text{ W/(m}^2\text{K)}$$

Now the heat transfer rate can be calculated:

$$\dot{Q} = h\pi dL(t_s - t_\infty) = 11947 \times \pi \times 0.006 \times 2 \times (35 - 15) = 6756 \text{ W}$$

13.6. *Assumption*: Steady-state conditions; temperatures are constant; no convective currents between plates.

The heat exchange between two finite gray bodies 1 and 2 at temperatures T_1 and T_2, respectively, ($T_2 > T_1$) is

$$\dot{Q} = \sigma\,\varepsilon_{12}\,F_{12}\,A\,(T_2^4 - T_1^4)$$

For two parallel walls, the emissivity factor ε_{12} is calculated as

$$\varepsilon_{12} = \frac{1}{\frac{1}{\varepsilon_1} + \frac{1}{\varepsilon_2} - 1} = \frac{1}{\frac{1}{0.75} + \frac{1}{0.65} - 1} = 0.534$$

and the view factor $F_{12} = 1$.

$$\frac{\dot{Q}}{A} = \sigma\,\varepsilon_{12}\,F_{12}(T_2^4 - T_1^4) = 5.67 \times 10^{-8} \times 0.534 \times 1 \times (850 - 600)$$

$$= 11887 \text{ W/m}^2 = 11.9 \text{ kW/m}^2$$

For black surfaces $\varepsilon_{12} = 1$. Therefore

$$\frac{\dot{Q}}{A} = \sigma(T_2^4 - T_1^4) = 5.67 \times 10^{-8} \times (850 - 600)$$

$$= 22249 \text{ W/m}^2 = 22.2 \text{ kW/m}^2$$

Bibliography

[1] G. Rogers and Y. Mayhew, *Engineering Thermodynamics: Work and Heat Transfer*, 4th edition, London: Prentice-Hall, 1992.

[2] "Thermodynamique — Wictionnaire," 28 July 2014. [Online]. Available: http:// fr.wiktionary.org/wiki/thermodynamique. [Accessed 4 August 2014].

[3] "Thermo-dynamics — Hmolpedia," 2012. [Online]. Available: http://www.eoht.info/page/Thermo-dynamics. [Accessed 4 May 2014].

[4] J. R. Taylor, *An Introduction to Error Analysis: The Study of Uncertainties in Physical Measurements*, Sausalito, CA: University Science Books, 1997.

[5] Bureau International des Poids et Mesures, BIPM — SI Brochure, 8th edition, 2006. [Online]. Available: http://www.bipm.org/utils/common/pdf/si_brochure_8_en.pdf. [Accessed 4 January 2015].

[6] A. A. Sonin, "The Physical Basis of Dimensional Analysis," 2001. [Online]. Available: http://web.mit.edu/2.25/www/pdf/DA_unified.pdf. [Accessed 4 January 2015].

[7] M. J. Moran, H. N. Shapiro, B. R. Munson and D. P. DeWitt, *Introduction to Thermal Systems Engineering: Thermodynamics, Fluid Mechanics, and Heat Transfer*, New York, NY: John Willey & Sons, Inc., 2003.

[8] D. N. Zubarev, "Thermodynamic Equilibrium," 2010. [Online]. Available: http://encyclopedia2.thefreedictionary.com/Thermodynamic+Equilibrium. [Accessed 4 January 2015].

[9] Z. S. Spakovszky, "1.2 Definitions and Fundamental Ideas of Thermodynamics," 2008. [Online]. Available: http://web.mit.edu/16.unified/www/FALL/thermodynamics/notes/node11.html. [Accessed 4 January 2015].

[10] E. Poisson, "Statistical Physics II (PHYS*4240) Lecture Notes (Fall 2000)," 2000. [Online]. Available: http://www.physics.uoguelph.ca/poisson/research/spii.pdf. [Accessed 5 May 2012].

[11] Y. A. Çengel and M. A. Boles, *Thermodynamics: An Engineering Approach*, 6th edition, New York, NY: McGraw-Hill, 2008.

[12] I. D. Morrison, "Single Component Phase Diagrams," 23 November 2011. [Online]. Available: http://soft-matter.seas.harvard.edu/index.php/Single_component_phase_diagrams. [Accessed 5 January 2015].
[13] J. Clark, "Phase Diagrams of Pure Substances," 2004. [Online]. Available: http://www.chemguide.co.uk/physical/phaseeqia/phasediags.html. [Accessed 6 January 2015].
[14] University of Florida, "Phases of Matter," [Online]. Available: http://itl.chem.ufl.edu/2045_s00/ lectures/lec_f.html. [Accessed 22 June 2011].
[15] E. W. Lemmon, M. L. Huber and M. O. McLinden, "NIST Reference Fluid Thermodynamic and Transport Properties — REFPROP Version 8.0; User's Guide," April 2007. [Online]. Available: www.nist.gov/srd/upload/REFPROP8.PDF. [Accessed 8 January 2015].
[16] E. I. du Pont de Nemours and Company, "Thermodynamic Properties of DuPont Suva 407C Refrigerant (R-407C)," 2004. [Online]. Available: http://www2.dupont.com/Refrigerants/en_US/assets/downloads/h56607_Suva407C_thermo_prop_si.pdf. [Accessed 7 January 2015].
[17] NIST Standard Reference Database 121, "Fundamental Physical Constants from NIST," January 2015. [Online]. Available: http://physics.nist.gov/cuu/Constants/. [Accessed 29 January 2015].
[18] C. Chieh, "The Ideal Gas Law," [Online]. Available: http://www.science.uwaterloo.ca/~cchieh/cact/c120/idealgas.html. [Accessed 10 January 2015].
[19] M. J. Blandamer and J. C. R. Reis, "Topic 1222 Equation of State: Real Gases: van der Waals and Other Equations," 2004. [Online]. Available: www.le.ac.uk/chemistry/thermodynamics/pdfs/1500/topic1222.pdf. [Accessed 11 January 2015].
[20] P. Subbarao, "Chapter 3 — Equations of State," [Online]. Available: http://web.iitd.ac.in/~pmvs/courses/mel140/EOS-vapor.pdf. [Accessed 11 January 2015].
[21] R. M. Price, "Real Gases — RMP Lecture Notes," 1996. [Online]. Available: http://facstaff.cbu.edu/rprice/lectures/realgas.html. [Accessed 11 January 2015].
[22] P. Nag, *Engineering Thermodynamics*, 4th edition, New Delhi: Tata McGraw-Hill, 2008.
[23] P. M. Bellan, "A Microscopic, Mechanical Derivation of the Adiabatic Gas Relation," 2004. [Online]. Available: http://ve4xm.caltech.edu/webpub/adiabatic-Bellan-AJP-2004.pdf. [Accessed 15 January 2015].
[24] D. P. Kolodnyi, "Generalized Definition of the Adiabatic Exponent," July 1965. [Online]. Available: http://link.springer.com/article/10.1007%2FBF00831838. [Accessed 15 January 2015].
[25] G. Creţa, *Tratat de turbine cu abur şi cu gaze*, (in Romanian), Bucureşti: Editura AGIR, 2011.
[26] R. Shanthini, "The First Law Applied to Steady Flow Processes," 12 January 2008. [Online]. Available: www.rshanthini.com/tmp/ThermoBook/ThermoChap10.pdf. [Accessed 16 January 2015].

[27] Z. S. Spakovszky, "4.3 Features of Reversible Processes," [Online]. Available: http://web.mit.edu/16.unified/www/FALL/thermodynamics/notes/node34.html. [Accessed 19 January 2015].

[28] D. L. Kormos-Buchwald, "Walther Hermann Nernst," Encyclopaedia Britannica, 2015. [Online]. Available: http://www.britannica.com/EBchecked/topic/409496/Walther-Hermann-Nernst/261432/Third-law-of-thermodynamics. [Accessed 20 January 2015].

[29] R. Serway and C. Vuille, *College Physics*, 9th edition, Cengage Learning, 2011.

[30] G. Elert, "Energy and Entropy — The Physics Hypertextbook," 2015. [Online]. Available: http://physics.info/thermo-second/. [Accessed 4 February 2015].

[31] Bureau International des Poids et Mesures, "SI Brochure: The International System of Units (SI) [8th edition, 2006; updated in 2014]," [Online]. Available: http://www.bipm.org/en/publications/si-brochure/section2-2-2.html. [Accessed 28 January 2015].

[32] Z. Morvay and D. Gvozdenac, "Supplementary Material — Fundamentals for Analysis and Calculation of Energy and Environmental Performance — Toolbox 6," November 2008. [Online]. Available: http://www.wiley.com/legacy/wileychi/morvayindustrial/supp.html. [Accessed 30 January 2015].

[33] Lulea University of Technology, "Mollier Diagram," 26 May 2005. [Online]. Available: http:// staff.www.ltu.se/~lassew/ene/MOLLIER.ppt. [Accessed 30 January 2015].

[34] P.-D. Oprisa-Stanescu and I. Ionel, "Ciclul HAT (The HAT Cycle — in Romanian)," Revista Termotehnica, pp. 47–52, No. 2, 2002.

[35] P. Gunter and M. Mariotte, "Hazards of Boiling Water Reactors in the United States," NIRS, March 1996. [Online]. Available: http://www.nirs.org/factsheets/bwrfact.htm. [Accessed 3 February 2015].

[36] R. Nave, "Typer of Nuclear Reactors," 19 February 2008. [Online]. Available: http://hyperphysics.phy-astr.gsu.edu/hbase/nucene/reactor.html. [Accessed 4 February 2015].

[37] L. Radulescu, "Contributions to the Study and the Research of Supercharged-Air Cooling at Four-Stroke CI Engines," Doctoral Thesis, "Politehnica" University, Timisoara, Romania, 1997.

[38] Asociación RUVID, "New Two-Stroke Engine, Notable for its Low Consumption and Low Level of Pollutant Emissions," Science Daily, 3 November 2014. [Online]. Available: http://www.sciencedaily.com/releases/2014/11/141103082518.htm. [Accessed 8 February 2015].

[39] Wikipedia Contributors, "Atkinson Cycle," 8 February 2015. [Online]. Available: http://en.wikipedia.org/wiki/Atkinson_cycle. [Accessed 8 February 2015].

[40] Image by User: Y_tambe – User: Y_tambe's file, CC BY-SA 3.0, https://commons.wikimedia.org/w/index.php?curid=290591. Available at https://commons.wikimedia.org/wiki/File:Wankel_Cycle_anim.gif (Accessed 6 August 2018).

[41] Wikipedia contributors, "Wankel Engine," The Columbia Electronic Encyclopedia, 8 October 2008. [Online]. Available: http://www.reference.com/browse/wankel+engine. [Accessed 8 February 2015].
[42] V. Ganesan, *Internal Combustion Engines*, New Delhi: Tata McGraw-Hill, 2008.
[43] Defense Technical Information Center, "Chapter 7 — Aero Propulsion (February 1991)," [Online]. Available: http://handle.dtic.mil/100.2/ada320215. [Accessed 11 February 2015].
[44] Institut International du Froid, "Classification of Refrigerants," 2001. [Online]. Available: www.iifiir.org/userfiles/file/webfiles/summaries/Refrigerant_classification_EN.pdf. [Accessed 14 February 2015].
[45] US Department of Energy, "Heat Pump Systems," 15 December 2014. [Online]. Available: http://energy.gov/energysaver/articles/heat-pump-systems. [Accessed 14 February 2015].
[46] Natural Resources Canada's Office of Energy Efficiency, "Heating and Cooling With a Heat Pump," December 2004. [Online]. Available: www.nrcan.gc.ca/sites/oee.nrcan.gc.ca/files/pdf/publications/infosource/pub/home/heating-heat-pump/booklet.pdf. [Accessed 14 February 2015].
[47] "The Speed of Sound," [Online]. Available: http://www.mathpages.com/home/kmath109/kmath109.htm. [Accessed 23 July 2012].
[48] World Energy Council, "World Energy Council Report Confirms Global Abundance of Energy Resources and Exposes Myth of Peak Oil," 15 October 2013. [Online]. Available: http://www.worldenergy.org/news-and-media/press-releases/world-energy-council-report-confirms-global-abundance-of-energy-resources-and-exposes-myth-of-peak-oil/. [Accessed 18 February 2015].
[49] World Coal Association, "Coal Statistics," 2013. [Online]. Available: http://www.worldcoal.org/resources/coal-statistics/. [Accessed 18 February 2015].
[50] Engineering ToolBox, "Optimal Combustion Processes — Fuels and Excess Air," [Online]. Available: http://www.engineeringtoolbox.com/fuels-combustion-efficiency-d_167.html. [Accessed 18 August 2012].
[51] The Greenhouse Gas Protocol , "Calculating CO_2 Emissions from Ammonia Production (Version 2.0)," 2006. [Online]. Available: www.ghgprotocol.org/files/ghgp/tools/Calculating%20CO2%20Emissions%20from%20Ammonia%20Production.pdf. [Accessed 18 February 2015].
[52] U.S. Environmental Protection Agency, Office of Air and Radiation, "The Climate Leaders Greenhouse Gas Inventory Protocol Core Module Guidance," May 2008. [Online]. Available: www.epa.gov/climateleaders/documents/resources/stationarycombustionguidance.pdf. [Accessed 18 February 2015].
[53] J. Ramage, *Energy: A Guidebook*, London: Oxford University Press, 1997.
[54] G. Goebel, "Fuel Cells," 1 July 2014. [Online]. Available: http://www.vectorsite.net/tpchem_13.html. [Accessed 19 February 2015].
[55] H. L. MacLean and L. B. Lave, "Evaluating Automobile Fuel/Propulsion System Technologies," *Progress in Energy and Combustion Science*, vol. 29, pp. 1–63, 2003.

[56] P.-D. Oprişa-Stănescu, *Autovehicule electrice, hibride şi cu pile de combustie*, Timişoara: Editura Politehnica, 2015.
[57] Toyota Europe Press Release, "Toyota Ushers in the Future with Launch of 'Mirai' Fuel Cell Sedan," Toyota Motor Corporation, 18 November 2014. [Online]. Available: http://newsroom.toyota.eu/newsrelease.do;jsessionid =03CA49F0A7CFB509D19E341B98CDF3DC?&id=4124&allImage=1&teas er=toyota-ushers-future-launch-mirai-fuel-cellsedan&mid=. [Accessed 8 March 2016].
[58] A. Bejan, *Heat Transfer*, John Wiley & Sons, Inc., 1993.
[59] Staff of Research and Education Association, Dr. M. Fogiel, Director, *The Heat Transfer Problem Solver*, New York, NY: Research and Education Association, 1986.
[60] D. Lelea and I. Laza, "The Particle Thermal Conductivity Influence of Nanofluids on Thermal Performance of the Microtubes," *International Communications in Heat and Mass Transfer*, no. 59, pp. 61–67, 2014.
[61] J. I. Lienhard and J. V. Lienhard, *A Heat Transfer Textbook*, Cambridge, MA: Phlogiston Press, 2008.
[62] Y. Çengel and A. J. Ghajar, *Heat and Mass Transfer (in SI Units)*, McGraw-Hill Education, 2014.
[63] K. Shaji, B. P. Rao, B. Sunden, W. Roetzel and S. K. Das, "Logarithmic Mean Pressure Difference — A New Concept in the Analysis of the Flow Distribution in Parallel Channels of Plate Heat Exchangers," *Heat Transfer Engineering*, vol. 33, no. 8, pp. 669–681, 2012.
[64] J. Kestin, *A Course in Thermodynamics*, vol. 1, Hemisphere Publishing Corporation, 1979.
[65] R. Fowler and E. A. Guggenheim, *Statistical Thermodynamics: A Version of Statistical Mechanics for Students of Physics and Chemistry*, Cambridge, UK: Cambridge University Press, 1939.
[66] R. B. Bird, W. E. Stewart and E. N. Lightfoot, *Transport Phenomena*, 2nd edition, New York: John Wiley & Sons, 2007.
[67] J. B. Tatum, "Chapter 6 — Properties of Gases," 2013. [Online]. Available: http://astrowww.phys.uvic.ca/~tatum/thermod/thermod06.pdf. [Accessed 11 January 2015].
[68] L. Vas, "Topics in Mathematics with Applications to Chemistry. State Functions," [Online]. Available: www.usciences.edu/~lvas/mathmethods/math 360/State_functions.pdf. [Accessed 12 January 2015].
[69] W. C. Carter, "Internal Energy of an Ideal Gas," MIT, 28 September 2002. [Online]. Available: http://pruffle.mit.edu/3.00/Lecture_11_web/node1.html. [Accessed 12 January 2015].
[70] F. Doumenc, "Elements de Thermodynamique et Thermique — I Thermodynamique," 2008–2009. [Online]. Available: www.fast.upsud.fr/~doumenc/ la200/CoursThermodynamique_L2.pdf. [Accessed 10 May 2012].

Index

Printed in the United States
By Bookmasters